# SmartCAM® Tutorials

John R. Johnson
Cape Fear Community College

Prentice Hall
*Upper Saddle River, New Jersey    Columbus, Ohio*

**Library of Congress Cataloging-in-Publication Data**

Johnson, John R.
    SmartCAM tutorials / John R. Johnson.
        p.    cm.
    Includes index.
    ISBN 0-13-685521-0
    1. Milling-machines—Numerical control—Software.  2. CAD/CAM
systems.  3. SmartCAM Production milling.   I. Title.
TJ1225.J645   1999
621.9' 1—DC21                                      98-6396
                                                                                   CIP

Editor: Ed Francis
Production Editor: Stephen C. Robb
Design Coordinator: Karrie M. Converse
Cover Designer: TopeGrafix
Production Manager: Patricia A. Tonneman
Production Supervision: Custom Editorial Productions, Inc.
Marketing Manager: Danny Hoyt

This book was set in Stone Serif by Custom Editorial Productions, Inc. and was printed and bound by Courier/Kendallville, Inc. The cover was printed by Phoenix Color Corp.

© 1999 by Prentice-Hall, Inc.
Simon & Schuster/A Viacom Company
Upper Saddle River, New Jersey 07458

All rights reserved. No part of this book may be reproduced, in any form or by any means, without permission in writing from the publisher.

SmartCAM® is a registered trademark of Structural Dynamics Research Corporation.

Printed in the United States of America

10 9 8 7 6 5 4 3 2 1

ISBN: 0-13-685521-0

Prentice-Hall International (UK) Limited, *London*
Prentice-Hall of Australia Pty. Limited, *Sydney*
Prentice-Hall Canada, Inc., *Toronto*
Prentice-Hall Hispanoamericana, S. A., *Mexico*
Prentice-Hall of India Private Limited, *New Delhi*
Prentice-Hall of Japan, Inc., *Tokyo*
Simon & Schuster Asia Pte. Ltd., *Singapore*
Editora Prentice-Hall do Brasil, Ltda., *Rio de Janeiro*

# PREFACE

In the seven-plus years during which I have taught SmartCAM production milling, I have searched in vain for a suitable book. To assist my students in comprehending what can be very deep subject matter, I began to write tutorials on a variety of projects. During those seven years, I compiled enough material to publish a text. I base my expertise in this subject matter on sixteen years of manufacturing experience. Thirteen of those years have been involved specifically with teaching machining technology, with the most recent seven years, as noted, teaching SmartCAM.

These tutorials are written for students who are involved in postsecondary education. In addition, students should ideally have experience in the Windows work environment, basic CNC programming, and machine tool operations. A basic knowledge of cutting tools, principles of metal removal, and material cutting speeds and feed rates is imperative for students to be successful with these tutorials. This text is not meant to take the place of formal classroom instruction. It is best used to supplement lectures and demonstrations conducted by an experienced user of SmartCAM.

The purpose of this text is to provide a step-by-step instruction manual on a wide range of two-dimensional sample workpieces that are representative of projects found in a modern manufacturing environment. This text is written for SmartCAM Production Milling version 9.2 and newer software. Each tutorial is written in a manner that will achieve a series of predefined objectives. As with anything else, however, there are several methods to accomplish any particular task with SmartCAM. After students begin to understand SmartCAM, they may discover simpler ways to achieve the same results that each tutorial demonstrates. In addition, students may need to modify the tooling and the cutting speeds and feed rates for the various workpiece materials used.

Although machine tool operation is not a part of this text, each tutorial is written partially from a manufacturing viewpoint. The idea is for students to build the part in SmartCAM and then actually machine the part on a CNC vertical machining center.

The text begins with simple, introductory projects in which basic foundation building skills are presented. Each tutorial then builds on skills that are learned in the previous tutorial. The final three tutorials demonstrate more advanced operations that require workpieces to be machined on all sides.

## Acknowledgments

For this project to come to fruition, many sacrifices were made by many people. I owe them an enormous debt of gratitude.

First and foremost, I would like to thank God for the many blessings He has bestowed upon me. Without His blessings this project would have never been possible. My prayer is that He would receive honor and glory for whatever outcome He may have for this text.

My darling wife, Cindy, and my daughter, Sarah, deserve many thanks for their understanding and the encouragement that they never failed to provide me during this year-long endeavor.

The efforts of my good friend, Joel Spencer, and his much-needed help with the blueprints will always be appreciated.

Last but not least, I thank the following reviewers who provided positive suggestions: Robert Brumm, Alfred State College–SUNY; John Gruben, Waubonsee College; James R. Hall, Macomb Community College; and Calvin Sams, Rowan-Cabarrus Community College.

# CONTENTS

**Chapter 1  The SmartCAM Environment  1**
    Title Bar  2
    Menu Bar  3
    Icon Bar  4
    Workbench  6
    Tool List  6
    Control Panel  9
    Database List  9
    Snap Icons  10
    Graphics Work Area  15
    Readout Line  15
    Modifications to Version 11 Software  16

**Chapter 2  Tutorial 1: The Job Operations File  19**

**Chapter 3  Tutorial 2: Slotted Plate  33**
    Creating the Finish Profile  44
    Constructing the Slot Profile  48
    Constructing the Roughing Profiles  55
    Constructing the Roughing Passes for the Slot  58

**Chapter 4  Tutorial 3: Pocket Mill/Drill  61**
    Constructing the Finish Profiles  63
    Adding Lead In/Lead Out Moves  68

Contents

    Construction of the Pocket Profile 69
    Constructing the Roughing Profiles 71
    Roughing the Pocket 75
    Drilling the Holes 78

## Chapter 5  Tutorial 4: Butterfly Flange  83

    Constructing the Finish Profiles 86
    Constructing the Pocket Finish Profile 92
    Constructing the "Island" 94
    Roughing the External Geometry 95
    Creating the Pocket Roughing Routine 98
    Constructing the Bolt Hole Circle 101
    Face Milling the Workpiece 104

## Chapter 6  Tutorial 5: Four-Hole Frame  107

    Constructing the Finish Profiles 109
    Constructing the Roughing Profiles 119
    Spot Drilling and Drilling the Holes 123
    Facing the Workpiece 126
    Construction of Workplanes 129

## Chapter 7  Tutorial 6: Support Bracket  135

    Creating the Finish Profile 137
    Creating the Roughing Profiles 144
    Machining the Pocket 146
    Alternate Pocket Roughing Routine 154
    Spot Drilling and Drilling the Holes 157
    Face Milling the Work 160
    Constructing Additional Workplanes 163

## Chapter 8  Tutorial 7: Idler Wheel Support  167

    Constructing the Roughing Cuts 176
    Machining the Pocket 178
    Spot Drilling and Drilling the Holes 186
    Machining the Slot 187
    Facing the Workpiece 190
    Constructing Additional Workplanes 192

## Index  197

# CHAPTER 1

# The SmartCAM Environment

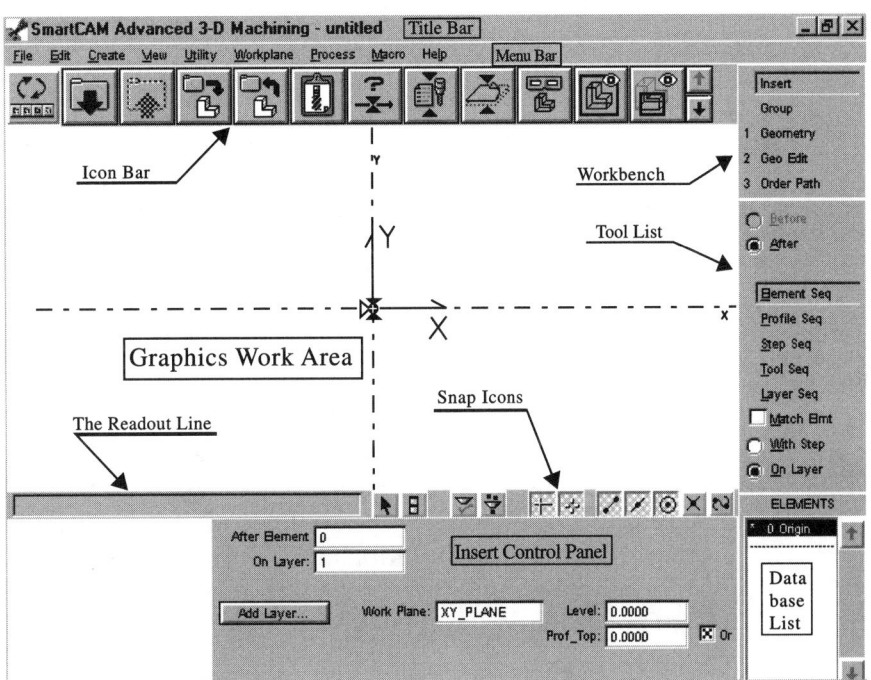

Upon completion of this chapter, you should be able to:

- Identify the areas of the SmartCAM environment.
- Understand the difference between a resident and a non-resident toolbox.
- Understand and apply the terminology associated with the SmartCAM environment.
- Correctly identify sub-menus, toolboxes, and dialog boxes by their markings from within the "pull down" menus.
- Read in and custom build a variety of icon bars.
- Understand the application of filters and correctly apply a filter to limit element selection.
- Identify an input field and correctly input data into an active input field.

# THE SMARTCAM ENVIRONMENT

SmartCAM Production Milling software is a very powerful and versatile computer aided manufacturing software package. With SmartCAM software, it is possible to manufacture a variety of work that may range from the most basic to the most complex machining operations. Additionally, SmartCAM software will allow you to design and manufacture parts either on a single-part basis, where you only work with one part on the screen at a time, or on a large volume, multi-part production basis, where many workpieces are shown simultaneously on the screen.

Before you can fully utilize SmartCAM, it is important to understand the parts of the SmartCAM environment. The SmartCAM environment is much like any other windows-based software environment. It consists of many menus, icons, control panels, and dialog boxes. As with many windows-based work environments, occasionally you will see certain selections shaded or "grayed out." An icon or selection that is "grayed out" indicates that this item is not a valid selection at this time. Several possibilities exist which would cause this condition. These possibilities will be discussed in depth as each situation presents itself. Additionally, as with other software programs, the more familiar you become with the specific areas and the menus within the SmartCAM environment, the easier it will be for you to manipulate the SmartCAM system.

The SmartCAM working environment consists of ten distinct areas: The title bar, the menu bar, the icon bar, the workbench, the tool list, the control panel, the database list, the snap icons, the graphics work area, and the readout line.

**Title Bar.** This area simply lists the name of the application's software and the particular file that you currently have open. In Figure 1.1, you will notice the word "untitled" in this area. This is because there is no active file in this illustration. As you read in a file, or as you save your file, this area will reflect

Chapter 1  The SmartCAM Environment

**FIGURE 1.1**
The SmartCAM environment

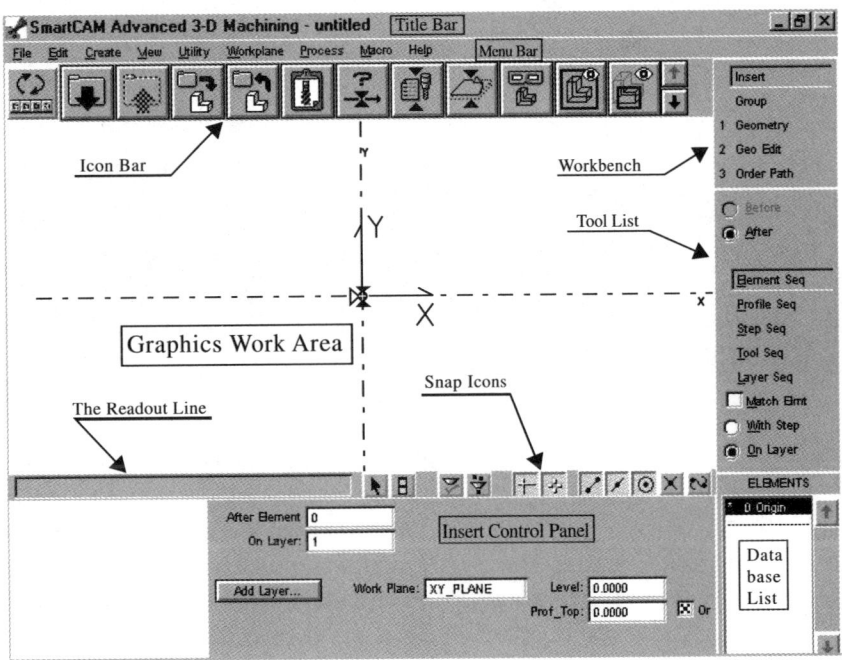

the title and path of your file. Take special note of this name (and path) so you can easily locate your file in the future.

**Menu Bar.**  Upon selection of one of these nine choices, SmartCAM activates "pull down" menus, each of which offers a variety of sub-menus, toolboxes, and dialog boxes.

Sub-menus are indicated by an arrowhead after the item name. *Submenus* require you to make additional choices in order to complete your objective. Figure 1.2 shows the selection of the **Utility** menu, along with the **Icon Bar** sub-menu.

**FIGURE 1.2**
The **Utility** menu and **Icon Bar** sub-menu

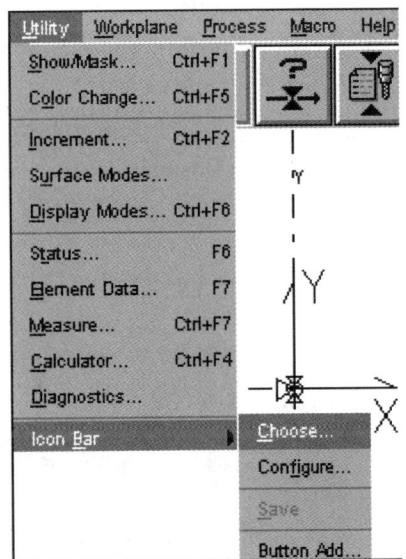

**4**  Chapter 1  The SmartCAM Environment

**FIGURE 1.3**
The **Create** menu

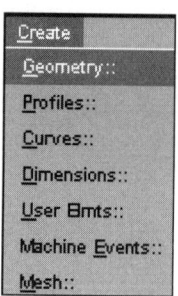

Dialog boxes are indicated by three periods after the item name. *Dialog boxes* are simply open windows that require data input from the operator. Figure 1.2 shows many selections that are followed by three periods. The selection of any one of these items will result in the opening of a dialog box.

Toolboxes are indicated by a square made of four dots after the item name. Upon choosing a toolbox from the menu bar, the specified toolbox will immediately go to the workbench. Figure 1.3 shows the selection of the **Create** menu. All selections within the **Create** menu are toolboxes.

Probably one of the most helpful menu selections SmartCAM offers to the new user is the **Help** menu. The Help menu is very comprehensive and very easy to use. Don't be afraid to explore the help screens if you run into a problem. Additionally, all SmartCAM user manuals are accessible from within the Help menu.

**Icon Bar.**  The icon bar offers a series of "hot keys" or shortcut keys for selections that are contained in the menu bar. The icon bar offers a much quicker route to these selections. There are several icon bars within the system that you can use.

To select a specific icon bar, follow these steps:

1. With your mouse, choose the word **Utility** from the menu bar.
2. Again with your mouse select **Icon Bar**.
3. Select **Choose**.
4. When the **Choose Icon Bar** dialog box opens, choose the **File Select** button.
5. Choose **Demo.bar**. (C:\sma\shared\ICON\Demo.bar)
6. Choose **Open**.
7. Choose **Accept**.

The **Demo.bar** icon bar is shown in Figure 1.4.

**FIGURE 1.4**
The **Demo.bar** icon bar

It is also possible to build and save a customized icon bar in SmartCAM. If you find yourself repeating a series of steps for a library of parts, it may benefit you to customize the icon bar to increase your productivity.

To custom build an icon bar, follow these steps:

1. Select **Utility**.
2. Select **Icon Bar**.
3. Select **Choose**.
4. When the **Choose Icon Bar** dialog box opens, select the file **Empty.BAR.** (This will load an icon bar that has no icons.)
5. Select **Utility.**
6. Select **Icon Bar.**
7. Select **Button Add**.

When the **Add Bar Button** dialog box opens, scroll through the selections under **Predefined Buttons**.

8. Select the word **Insert**.
9. Select **Accept**.

The **Insert** icon is now added to the icon bar.

10. Again select **Utility > Icon Bar > Button Add**.
11. Under **Predefined Buttons** select the word **Group**.
12. Select **Accept**. The **Group** icon is added to the icon bar.

Continue the above procedure until you have the **Insert** icon, the **Group** icon, the **Geometry** icon, and the **Geo Edit** icon on the icon bar as shown in Figure 1.5.

It is also possible to delete icons from any of the icon bars. To delete an icon that you no longer need, select the icon with your mouse while holding down the shift key of your keyboard. This will open the **Edit Bar Button** dialog box. When the window opens, simply choose the **Delete** button as shown in Figure 1.6.

Once you have an icon bar customized to your satisfaction, you may want to save it.

To save a customized icon bar, follow these steps:

1. Choose **Utility** from the menu bar.
2. Choose **Icon Bar**.
3. Choose **Save**.

**FIGURE 1.5**
The **Insert, Group, Geometry** and **Geo Edit** icons

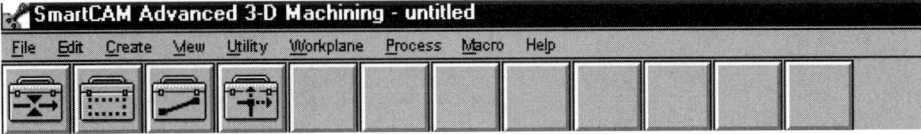

**FIGURE 1.6**
The **Edit Bar Button** dialog box

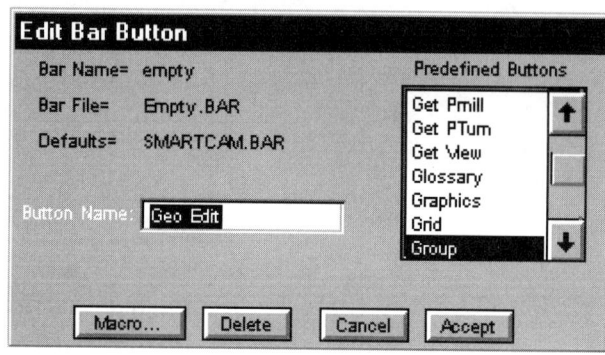

**Workbench.** The workbench holds five items that are referred to as toolboxes. When a toolbox is selected, it opens and reveals a series of specific modeling tools, the names of which are displayed in the tool list area.

Two of the toolboxes, the **Insert** toolbox and the **Group** toolbox, are referred to as workbench resident toolboxes. A *workbench resident toolbox* is one that remains on the workbench permanently. During the construction of a workpiece, these two toolboxes are used very frequently. It would therefore be antiproductive to constantly choose these two toolboxes from the menu bar in order to construct a workpiece in SmartCAM. For that reason they constantly reside on the workbench. You cannot remove them.

The other three toolboxes that reside on the workbench are non-resident. They are automatically replaced by the system depending upon the frequency of their use. For example, if a new toolbox is chosen from either the menu bar or the icon bar, the toolbox on the workbench that has been closed for the longest length of time is replaced by the new toolbox.

If you need to use a modeling tool from a particular toolbox that is on the workbench, simply pick that toolbox with your mouse. It will open and reveal the tools available to you. If the toolbox you need to use is not on the workbench, choose it from either the menu bar or the icon bar. It will automatically show up on the workbench and replace the toolbox that has been closed the longest. There will never be any more or any less than five toolboxes on the workbench as shown in Figure 1.7.

**Tool List.** This area lists the modeling tools that are available to you once you have opened a particular toolbox. The open toolbox, as well as selections from

**FIGURE 1.7**
The Workbench of SmartCAM

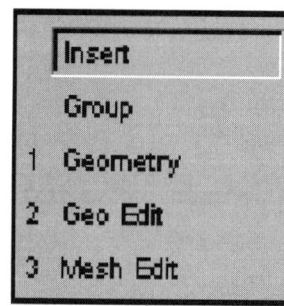

Chapter 1  The SmartCAM Environment        7

**FIGURE 1.8**
The **Insert** toolbox

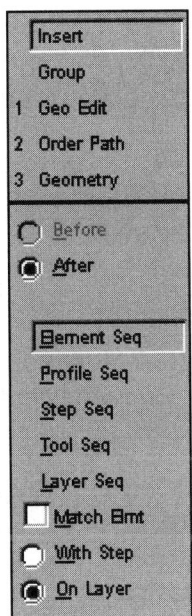

the tool list, will appear as a button that has been depressed. *Modeling tools* are tools used to create and properly place geometry.

Figure 1.8 shows that the **Insert** toolbox is open. The tool list reveals the modeling tools that are available in the **Insert** toolbox. The **After** selection has also been chosen. Notice the filled-in dot beside the item name. This indicates a selected item. **Element Seq** has been selected, as well as **On Layer.**

It may benefit you at this time to discuss the concept of grouping in SmartCAM. You have already seen a Group toolbox on your workbench. Next, a few icons will be discussed that relate to the concept of grouping. *Grouping*, in SmartCAM, is the process of identifying certain elements with which you wish to work. When you choose those items, they are said to be in the active group. Many modeling tools of SmartCAM require an active group before they can be used.

Upon selecting the **Group** toolbox from the workbench, the toolbox will open to reveal the modeling tools shown in Figure 1.9.

**FIGURE 1.9**
The modeling tools within the **Group** toolbox

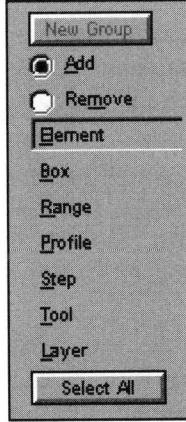

**New Group** will degroup an active group of elements. It is often necessary to degroup the elements that are in the active group prior to defining a new group of elements. Figure 1.9 shows the **New Group** selection grayed out. This selection will only be available once a group is active.

**Add** and **Remove** are selector switches that must be set prior to choosing any of the listed modeling tools. If there is no active group, the setting will default to **Add**. Once there is an active group, you will have the choice of adding elements to the active group or removing elements from the active group.

**Element** allows the selection of single elements using the mouse. The desired elements may be selected from either the graphics work area or the database list. Additionally, the number of the element may be entered into the **Group Element** control panel at the bottom of the screen. Care must be taken, however, when elements are chosen from the graphics work area. As the process model becomes more detailed and complex, it is often difficult to choose the correct element due to the congestion.

**Box** allows the selection of multiple elements by defining a region or box on the screen with two corner locations. Upon choosing this modeling tool, a control panel will open and prompt you for the choice of **Box Complete** or **Box Partial. Box Complete** selects those elements that reside entirely within the box. **Box Partial** selects those elements in which any portion of the element lies within the box. The **Box** option will choose all elements within the region regardless of their "Z" level.

To use **Box**, choose the **Box** modeling tool from the tool list and move the cursor to the graphics work area. Depress and hold the left mouse button and drag the mouse away from the starting location. When the box encompasses the desired elements, release the mouse button and all elements will be selected.

**Range** allows the selection of elements by specifying an element starting number and an element ending number. Upon selecting this modeling tool, a control panel will open and require input. Valid input consists of selections from the graphics work area, selections from the database list, or numerical entries from the keyboard.

**Profile** allows the selection of all elements that make up a profile. A *profile* is defined as a series of elements that are in sequential order in the database and adjacent in the model. As with the other modeling tools, you have the option to choose the profile from the graphics work area or the database list. Additionally, upon choosing **Profile**, the **Group Profile** control panel will open at the bottom of the screen and allow you to enter the number of an element that is part of the profile.

**Step** allows the selection of all elements that are assigned to the selected step. You may also choose the step from the graphics work area, from the database list, or by entering the step number into the **Group Step** control panel.

**Tool** allows the selection of all elements that are assigned to the selected tool. Selection of the tool may be from the database list, the graphics work area, or by entering the number of the tool into the **Group Tool** control panel.

**Layer** allows the selection of all elements that are assigned to the selected layer. You may also choose the layer by any of the three methods that have been previously described.

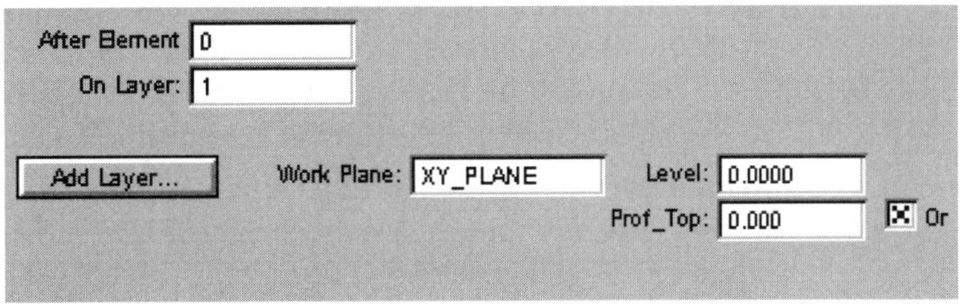

**FIGURE 1.10**
The **Insert** control panel

**Control Panel.** Upon making the applicable choices from the tool list, a control panel will open at the bottom of the SmartCAM screen. The control panel contains *input fields*, or areas in which data must be entered. This data will allow you to properly construct geometry based on your toolbox and modeling tool selections.

Figure 1.10 shows the **Insert** control panel that, upon selecting **Insert** from the workbench, will open at the bottom of the screen. The number 0 has been entered into the **After Element** input field. This arranges the database so that the next element that is created will be entered into the database immediately after element 0. The geometry will be assigned to layer 1, as per the **On Layer** input field, and the geometry will reside on the "XY" workplane. It will be constructed at a "Z" level of 0.00", and the top of the profile will reside at 0.00".

These values are for example's sake only. More realistic values will be shown in the following tutorials.

**Database List.** This area of the SmartCAM environment lists all data that you are building into your database. It will also list all elements that you are creating in the construction of your process model (your workpiece). Additionally, it will list all tools, steps, and layers that you are using to construct your geometry. Finally, it will list all defined workplanes that exist in your file. Figure 1.11, however, reveals a database that has no data. No file has been read in, nor has any geometry been created. This is characteristic of a newly opened file that you are just beginning. The tutorials of later chapters will allow you to watch your database grow as you construct geometry.

Notice the dotted line that is shown in the database list. The placement of this dotted line is controlled by your **Insert** toolbox and your **Insert** control panel. This dotted line indicates the insertion point of your next piece of geometry. You can manipulate the placement of this line by the **Before** or

**FIGURE 1.11**
The **Database list** of Smart-CAM

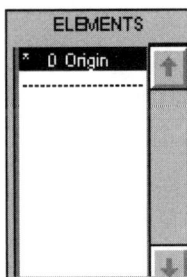

**After** selection of the **Insert** toolbox and by the data input into the various input fields of the **Insert** control panel.

The organization of the database and the ability to manipulate the data in the database is among the more important concepts of SmartCAM.

**Remember:** *It is very important for you to understand that the order in which the workpiece is machined at the machine tool is directly controlled by the order of the database.*

If the database is out of sequence, the workpiece will be machined out of sequence. Out of sequence machining will result in a crash at the machine tool. The tutorials that are included in this text will allow you to practice "database management."

**Snap Icons.** The snap icon area, located above the control panel, is a new feature that was introduced in version 9.2. It's an area that is actually divided into three smaller areas. The first area is made up of two icons which enhance the group toolbox. Next, are two icons that allow the filtering of elements. Finally, there is an area of seven icons that will enhance the ability to either snap to existing geometry or to free pick certain locations.

This icon will allow you to add existing elements to the active group. Simply pick the icon with the mouse and then pick the elements you wish to group. Element selection is controlled by settings contained in the edit filter icon and the group tool palette icon.

This is the **Group tool palette.** Choose this button with your mouse, and the tool palette will open revealing more icons.

The icons in Figure 1.12 work exactly as their modeling tool counterparts of the tool list work. However, the tool palette offers the opportunity to more effectively utilize the grouping function in conjunction with the edit filters to maximize the capabilities of SmartCAM. Additionally, there are a few added features of the tool palette which make it more desirable to use.

- Holding the control button while clicking on an element, step, or layer will remove it from the active group.
- With the Group by element selected, double-clicking an element in a profile adds it to the group.
- With either the layer or step option selected, triple-clicking any element associated with the desired step or layer adds it to the active group.

**FIGURE 1.12**
The icons located in the **Group tool palette**

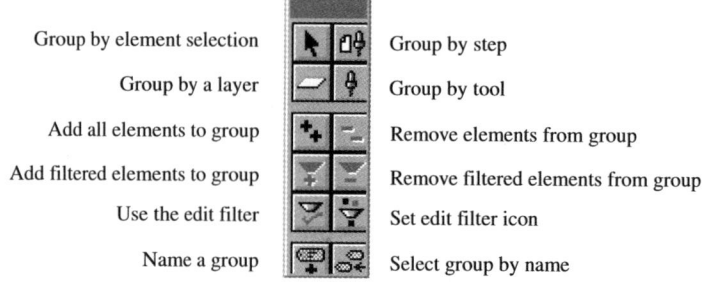

Chapter 1    The SmartCAM Environment

- A range of elements can easily be added to the active group by selecting the arrow from the group tool palette, selecting the first element in the range from the database list and, while holding the shift key, selecting the last element in the range.
- With the Group by element selected, double-clicking the mouse in a section of the graphics work area where there are no elements will add all elements to the active group.

To group by a box, follow these steps:

1. Select the **Group by element** from the tool palette.
2. Move the mouse to an area of the graphics work area.
3. Depress the mouse button and drag the box across the graphics work area.

All elements partially in the box are selected. To select elements completely in the box, follow the previous instructions while holding the shift key.

This is the **Edit Filter** icon. Upon selection of this icon, the Edit Filter dialog box will open. The Edit Filter icon is accessible from either the Group tool palette or the Snap icon line area above the control panel. Use the Edit Filter to set the selection filtering criteria when you select an active group, view element data, or use Snap mode to enter coordinate values. When you use the filter, the group, snap, and element data will only recognize the element types that you specify in the Edit Filter dialog box. To specify the element types simply choose the appropriate button from the Edit Filter dialog box. A depressed button signifies the selection is on, while a released button signifies the selection is off.

To demonstrate the function of the **Edit Filter** you must first read in a file. To read in a file, follow these steps:

1. Select **File** from the menu bar.
2. Select **Open**.
3. When the dialog box opens, select the file titled **Geopract.pm4**. (The full path is C:\SM9\MILL\MDATA\GEOPRACT.PM4) Figure 1.13 shows Geopract.pm4.
4. Choose **Accept**.
5.  Open the **Edit Filter** dialog box by choosing the shown icon. Figure 1.14 shows the Edit Filter dialog box.
6. Filter out all geometry types by selecting the **None** button.
7. Select the **Accept** button.
8.  To enable the use of the filters, select the **Use Edit Filter** icon from the snap icon line.
9. Select **Group** from the workbench.
10. Select **Box** from the tool list.
11. Create a box around the geometry. To create a box around the geometry, follow these steps:
    A. Move the mouse to the upper left corner of the geometry.

**FIGURE 1.13**
Geopract.pm4

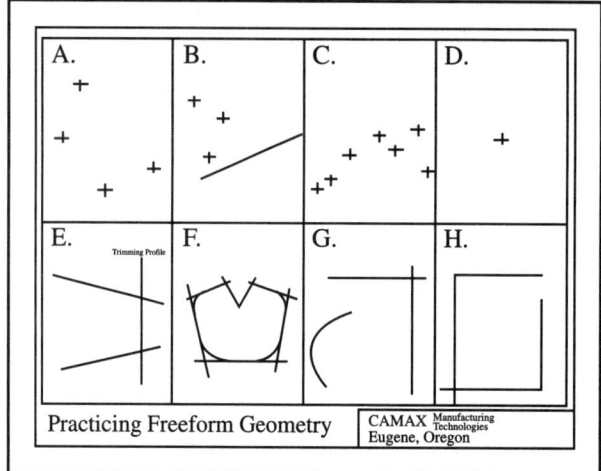

**FIGURE 1.14**
The **Edit Filter** dialog box

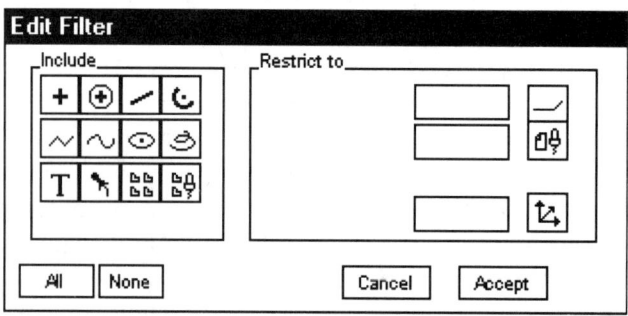

B. Depress and hold the left mouse button and move the mouse to the lower right corner of the geometry.

C. Release the mouse button.

Notice that no geometry was chosen because you previously filtered out all elements.

The best way to understand the idea of filtering is to visualize element types actually passing through the filter. Since you chose the **None** button, no elements were allowed to pass through the filter. Therefore, no elements were chosen.

12. Select the shown icon again to reopen the **Edit Filter** dialog box.
13. Select the **All** button.
14. Select **Accept**.
15. Repeat steps 9 through 11.

Notice this time that all elements within the box were grouped. This is because the settings in the Edit Filter dialog box allowed the selection of all elements. No elements were filtered out. All elements were allowed to pass through the filter.

This is the **Free Coordinate Mode** icon. **Free Coordinate Mode** is usable only in the **Create** and the **Edit** toolboxes. When you depress the

Chapter 1 The SmartCAM Environment 13

Free Coordinate Mode icon and activate it, the cursor is displayed as large crosshairs. As you move the crosshairs across the graphic view, the coordinates are displayed on the readout line. If you are attempting to enter coordinate values into an input field, simply move the cursor to the correct value as displayed in the readout line, and click the left mouse button. The values will be entered into the highlighted input field. The **Linear Increment** setting, under **Utility > Increment**, allows the setting of the incremental movement of the cursor.

This is the **Snap Mode** icon. When **Snap Mode** is on, the cursor is displayed as short crosshairs and, as usual, the icon will appear to be pressed in. **Snap Mode** is very helpful during the construction of a process model. Often, you will find it necessary to enter data into various control panel input fields by selecting elements from the graphics work area with your mouse. **Snap Mode** will assist you, in that when you position your mouse close to the element and click the left mouse button, the cursor will jump to or "snap to" the element. Your mouse does not need to be exactly on the element. As long as your cursor is within the range specified in the Increment dialog box the snap action will take place.

When **Snap Mode** is active, you have the option of five different settings that will determine the location of the snap point. The first of these icons—and probably the most frequently used—is the **End Point** icon. The **End Point** icon will allow you to snap to or select the start points and the end points of elements that already exist in the graphics work area. The **Mid Point** icon allows you to snap to the midpoint of all elements in the process model, except for polylines and splines. The **Center Point** icon allows you to snap to the center of all arc elements in the process model. The **Intersection** icon allows you to snap to the intersection points of lines or arcs that are in your database. **Intersection** does not recognize the intersection of polylines and splines.

The last icon is the **Control Point** icon. The **Control Point** icon is used to snap to the control points that make up polylines and splines.

SmartCAM versions 9.2 and newer also give you the option of setting your snap icons to Automatic Mode. Automatic Mode requires that you activate the **Snap Mode** icon and the **Free Coordinates Mode** icon simultaneously. The system switches between modes depending on the distance of the cursor from the nearest available point (the selection distance is set in the **Increment** dialog box). When the cursor is within the selection limit of an existing point, snap mode is in effect. Otherwise, the free coordinate mode is in effect. During automatic mode, you are only able to snap to those element types that are indicated by the Snap Point Settings icons.

**Hint:** *The Snap mode icons are restricted by the settings in the Edit Filter dialog box.*

For example, if your process model consists entirely of lines, and your Edit Filter dialog box is set to restrict the selection of lines, you will not be able to snap to those lines regardless of which snap icon is active.

To illustrate the functions of the snap icons and the combined function of the snap mode and the filters with automatic mode in effect, follow these steps:

1. Open the **Edit Filter** dialog box by selecting the shown icon.
2. Select **None**.

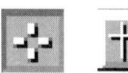

3. Select the shown icon to allow the selecting of point geometry.
4. Select **Accept**.
5. From the snap icon bar, select the shown icons to activate Automatic mode.
6. Select **Insert** from the workbench.
7. Select **After** from the tool list.
8. Select **Element Seq** from the tool list.
9. Select **Match Element**. (An "X" in the box beside Match Element indicates the "on" condition.)
10. Select **Geometry** from the workbench.
11. Choose **Line** from the tool list.
12. Move your mouse cursor to the word **Start Point** in the **Line** control panel at the bottom of the screen and select it with your mouse. It should turn white, if it is not already. A white input field means that it is active.
13. Move your mouse cursor close to the start point as shown in Figure 1.15 and depress the left mouse button. Notice how the mouse cursor switches from the Free Coordinate Mode to the Snap Mode as you near the start point.
14. Move your mouse cursor close to the end point as shown in Figure 1.15 and depress the left mouse button.

A line should be created as shown in Figure 1.16.

**FIGURE 1.15**
The start and end points

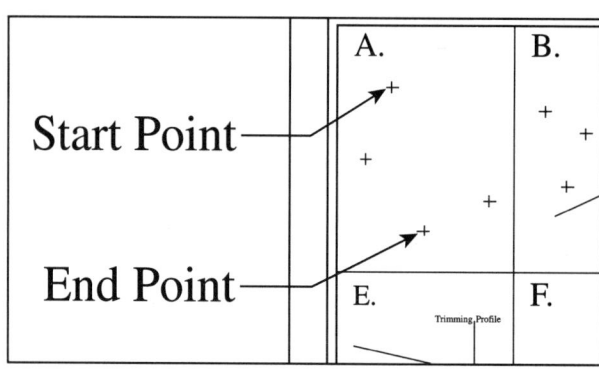

**FIGURE 1.16**
The line connecting the start and end points

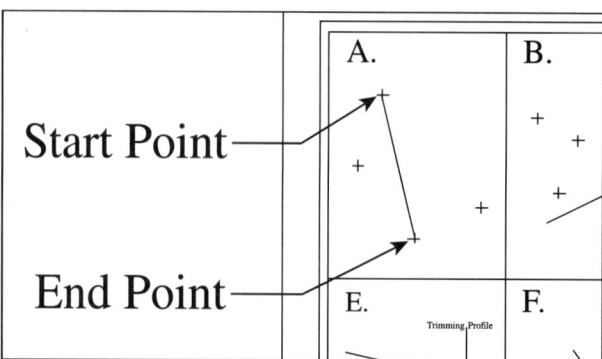

Chapter 1 The SmartCAM Environment

Continue by attempting to select the endpoint of one of the lines. To do this, follow these steps:

1. Refer to the **Line** control panel at the bottom of the screen to verify that the word **Start Point** is white. An input field that is white is considered active and ready to receive input. If it is not white, simply select the word **Start Point** with your mouse.

2. Turn the **Automatic Mode** off by reselecting the icon. This will limit the selections to the snap mode. Snap mode is needed for this demonstration.
3. Choose the starting and ending points for the next line according to Figure 1.17:

Notice that SmartCAM will not allow you to complete this task because the ending point is a line element. Previously, you filtered out the line elements. Therefore, you cannot choose them until you turn those element types on in the Edit filter dialog box.

**Graphics Work Area.** The *Graphics work area* is the working area in which you build your process model. The appearance of the Graphics work area is controlled by the settings in the **Display Modes** dialog box (**Utility > Display Modes**). You may want to experiment with the **Work Plane Indicator** selection, the **Grid** selection, the **World XYZ Axes** setting, and the **Rulers** setting to determine your preference. In regard to the selector switches within all areas of SmartCAM, an "X" in the box signifies the option is turned on while an empty box signifies an off condition.

**Readout Line.** The readout line is always visible below and to the left of the Graphics Work Area. The information displayed in this area will change to match the task that is currently in progress within SmartCAM. As you move the cursor over any menu selection, icon, toolbox, or tool, the readout line will display a short statement that identifies the item on which the cursor is resting. When the Free Coordinate mode is active the readout line will display the position of the cursor. Additionally, during Snap mode, the readout line will display the snap position coordinates of the cursor.

**FIGURE 1.17**

Start and end points, part B

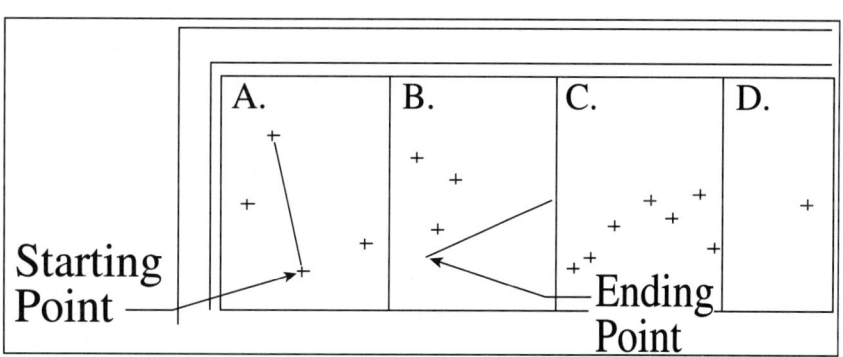

# Modifications to Version 11 Software

During the final phases of the completion of this text, Structural Dynamics Research Corporation released SmartCAM version 11 Production Milling. Many changes to the SmartCAM environment were made in version 11. If you are using version 11 software, many of these changes will affect this tutorial and need to be discussed before proceeding.

Among the most notable of these changes is the reconstruction of the workbench. The two workbench resident toolboxes, **Insert** and **Group**, have both been removed from the workbench.

The word **Insert,** as well as all tools associated with the **Insert** toolbox, have been replaced by the **Insert Property Bar**. The **Insert Property Bar** is made up of five major icons—two of which contain input fields and four additional input fields. Figure 1.18 shows the **Insert Property Bar**.

If you are using SmartCAM version 11 software, you will use this icon bar anytime you are given instructions concerning the **Insert** toolbox, any tool associated with the **Insert** toolbox, or the **Insert** control panel.

The first icon of the **Insert Property Bar** is the **Sequence Position Icon.** This icon has two settings and allows you to control the sequence of the database.

This is the Insert Before Setting of the Sequence Position Icon.

This is the Insert After Setting of the Sequence Position Icon. (Notice the position of the hour glass.)

These icons take the place of the **Before** and **After** selector switches that were found in the **Insert** tool list of versions 9 and 10.

Once you have determined the sequence of the geometry you are about to insert, you must then select the geometry sequence type.

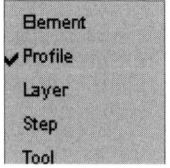
The second icon of the **Insert Property Bar** is the **Sequence Type Icon**. It is one of the two icons which has an input field. Upon selecting this icon, a menu opens which lists the geometry types :

Once the **Sequence Type** has been selected, the number of the type is entered into the corresponding input field. Figure 1.19 shows a line element has been chosen. Eighteen has been entered into the input field to indicate element number 18.

**FIGURE 1.18**
The **Insert Property Bar**

**FIGURE 1.19**
A chosen line element

The third icon of the **Insert Property Bar** is the **Match Properties** on/off switch.

Match properties is off when the icon is shown with a slash through it. When match properties is off, you will need to enter appropriate values into the input fields of the remaining icons to meet the requirements of your process model.

Match properties is on when the icon is shown as a depressed button with no slash. When match properties is on, the values in the input fields of the remaining icons are automatically set. Those input fields will match the properties of the geometry type that was chosen during the sequence type selection as was shown and discussed previously.

The fourth icon of the Insert Property Bar allows you to specify construction of geometry either with a step or on a layer. It is the second icon which contains an input field.

When the icon is represented in this manner, geometry will be assigned to a step.

When the icon is represented in this manner, geometry can be constructed on a layer.

Upon selecting the icon, a menu will open, allowing you to select the appropriate choices for the construction of your process model as shown in Figure 1.20.

This menu will also facilitate the construction of a job plan "on the fly" as you will see in tutorial number six.

The input field of the With Step/On Layer icon is found immediately after the icon. This input field allows you to specify the step number or the layer number on which you wish to create the geometry.

If the choice for the geometry construction type is a step, you must then specify the offset that is to be applied to that step. The fifth icon of the **Insert Property Bar,** the **tool offset icon,** allows you to specify the correct offset.

This icon of the **Insert Property Bar** indicates a step with no offset.

This icon indicates a step with a left offset.

This icon indicates a step with a right offset.

If the choice for the geometry construction is a layer, this icon will not be shown on the **Insert Property Bar**.

The remaining areas of the **Insert Property Bar** are input fields which require input from the operator.

This is the level input field of the **Insert Property Bar**. This input field allows you to set the "Z" level of your geometry. Additionally, selecting the arrow will display a menu that reveals a data list of the previous five entries.

**FIGURE 1.20**
The With Step/On Layer menu

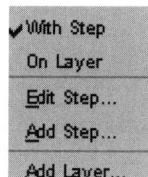

This input field allows you to specify the profile top value for your geometry. Again, by selecting the arrow, the last five values that were entered as profile top values are available for selection.

 This input field allows the selection of the clearance plane for geometry assigned to a step. It is not available for layer geometry. As with the previous two input fields, selecting the arrow will open a menu, revealing the five most recent entries.

This input field allows you to select a work plane on which your geometry is placed.

Another feature found in SmartCAM version 11 is the ability to alter the size and location of—and in some cases, completely hide—many of the icon bars. Version 11 gives you this flexibility with the **Insert Property Bar**. To hide or alter the size and location of the **Insert Property Bar,** follow these steps:

1. Choose **Utility** from the menu bar.
2. Select **Insert Property Bar.**
3. A menu will open as follows:

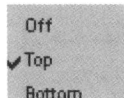

This menu gives you the ability to turn the **Insert Property Bar** off, move it to the top of your screen, or to the bottom of your screen.

As was stated earlier, SmartCAM version 11 also removed the word **Group** from the workbench. All grouping functions have been replaced by the Group Tool Palette. The appearance and function of the **Group Tool Palette** is exactly like that in version 10.

Many other changes were made to version 11 and were not specifically stated in this chapter. To find more information on a particular topic, refer to the on-line help screen and the on-line manuals supplied with the software.

CHAPTER 2

# SmartCAM Tutorial 1

## The Job Operations File

Upon completion of this chapter, you should be able to:

- Identify the two methods used to create geometry in SmartCAM and describe an application of each.
- Define the purpose of the Job Operations Planner.
- Correctly introduce a valid job file into a process model by each of the three methods.
- Identify the areas of the Job Operations Planner.
- Correctly enter data into the input fields of the Job Operations Planner.
- Read in a "pre-existing" job file and edit it to meet your needs.
- Recognize potential errors in modifying job files and take appropriate steps to correct those errors.

# THE JOB OPERATIONS FILE

Geometry construction within the SmartCAM environment can be accomplished in one of two ways: on a layer and with a step. Geometry created on a layer is done so to represent "non-toolpath" geometry. Anything you do not wish to machine should be created on a layer. Examples of this may be a fixture or a vise that locates or holds your workpiece. This geometry may be necessary to show the correct relationship between toolpath and workholder. Also, it may occasionally be necessary to show the material blank so you can position multiple workpieces for maximum efficiency. In addition, SmartCAM has a feature called Show Cut, which is a three-dimensional solid model representation of the movement of the cutting tool that utilizes geometry created on a layer to represent the workpiece.

The second method of creating geometry in SmartCAM is to create it with a Step. Creating geometry with a step means that any geometry that is created will be associated with a cutting tool and therefore will have "G" code properties. The principle of the step is to give the operator independent control of the speeds and feeds of the same cutting tool when that tool is used in different applications. For example, you may wish to use the same roughing tool in two different applications. The first application requires roughing the perimeter of a workpiece. The small amount of material engagement along the perimeter may allow you to increase your speeds and feeds. However, the second application requires a full engagement of the cutting tool, such as in a pocket. You would therefore need to decrease your feed rate to avoid a crash. Simply associating the same cutting tool with a different step gives you independent control of the speeds and feeds of that tool in situations just described. Prior to SmartCAM version 9 milling software, the only way to change the speeds and feeds of a cutting tool was to actually change the cutting tool itself. This required additional cutting tools, additional slots in the machine tool carousel, and time wasted during unnecessary tool changes. With the introduction of the step in SmartCAM version 9, and newer software, this is no longer necessary.

# Chapter 2  SmartCAM Tutorial 1

The area of the SmartCAM environment that allows you to select and properly set up the steps and the tooling that is associated with your steps is called the *Job Operation Planner*. It is one of the most important areas of SmartCAM. The Job Operation Planner allows you to specify a variety of information concerning the physical properties of the cutting tools you will use. Additionally, the information concerning the operation of those cutting tools can also be specified. Once the information contained in the Job Operation Planner is saved, it becomes your Job Operations File (JOF). These files will have a ".jof" extension.

**Remember:** *An active and valid Job Operations File is necessary before you can generate "G" code with SmartCAM.*

There are three ways to introduce a valid JOF into your process model:

1. A Job File that has been previously created and saved can be used. To use a previously saved Job File, simply choose **File** > **Load Job File** > **File Select** and choose the appropriate file name.
2. A Job File can be created, as a preliminary step, specifically for the process model on which you are actively working. This can be accomplished by choosing **File** > **Planner** or by choosing the **Job Planner** icon from the icon tool bar.
3. A Job File can be created "on the fly" or as the need arises for a specific tool or step. This can be accomplished by choosing the **With Step** tool from within the **Insert** toolbox. However, upon initially choosing **With Step**, the error message shown in Figure 2.1 will be displayed.

By choosing "OK" the Job Planner will automatically open. At this point, the completion of the input fields is the same as with other methods and will be demonstrated in this tutorial. All three methods will be demonstrated periodically throughout this text.

As stated earlier, when you save the file that you created in the Job Operation Planner, SmartCAM automatically assigns a ".jof" (job operations file) extension to your file. Older versions of SmartCAM (prior to version 9 software) assigned a ".jsf" extension to the tooling file. The ".jsf" extension indicates that the file is a Job Setup File. In working with SmartCAM, you will most likely encounter many files with the ".jsf" extension. Furthermore, you will occasionally need to read in a ".jsf" file into your version 9 software. It is a simple matter that is accomplished by setting the JSF conversion to the "on" position when prompted to load an existing job file, as shown in Figure 2.2.

By choosing **Accept** the existing Job Setup File, "Sample.jsf," will be converted to the JOF format for use in the version 9 and newer software.

**FIGURE 2.1**
The **With Step** error message

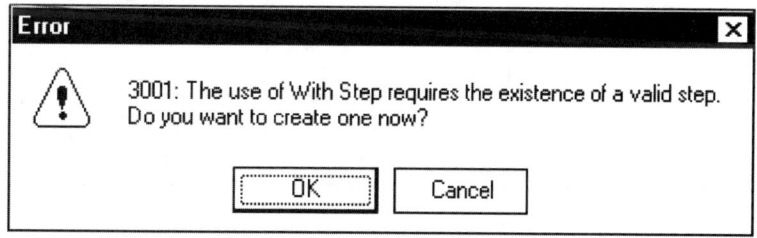

**FIGURE 2.2**
The JSF conversion in the "on" position

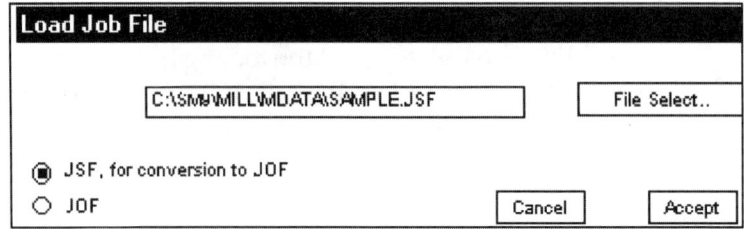

As stated earlier, a job file can be created as a preliminary step, prior to building geometry. To demonstrate this procedure, follow these steps:

1. Open SmartCAM.
2. Select **File.**
3. Select **Planner.**

An alternate method is to choose the **Process Planner Icon** from the icon tool bar. When the Job Operation Planner opens, choose the **Job Info** button as shown in Figure 2.3.

The job information area will allow you to list information about your specific process in three categories as shown in Figure 2.4.

**General** allows you to document the creator of the process model, type a brief description of the material blank, choose the units of measure, and write a brief description of the job.

**Machine** allows you to specify the machine ".smf" file and the template file which will be used to create a machine code once your process model is complete.

**FIGURE 2.3**
The **Job Info** button of the Job Operation Planner

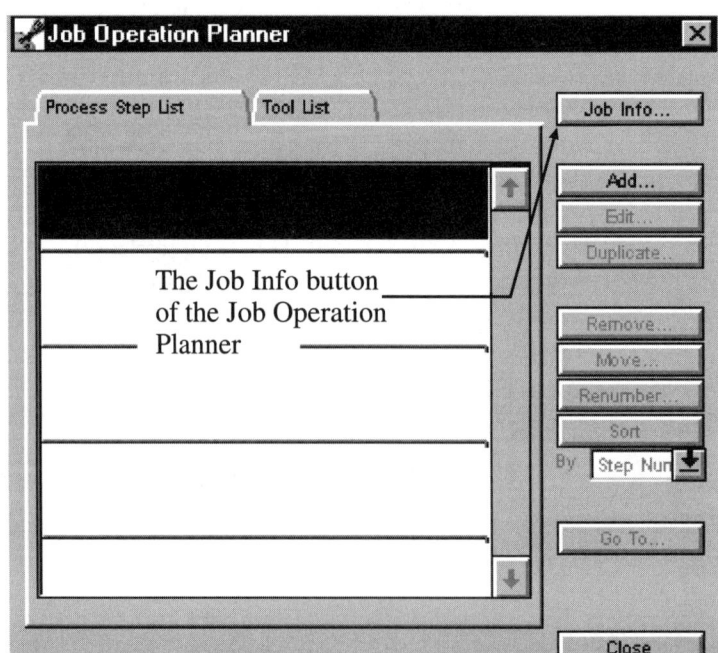

Chapter 2  SmartCAM Tutorial 1

**FIGURE 2.4**
The information page

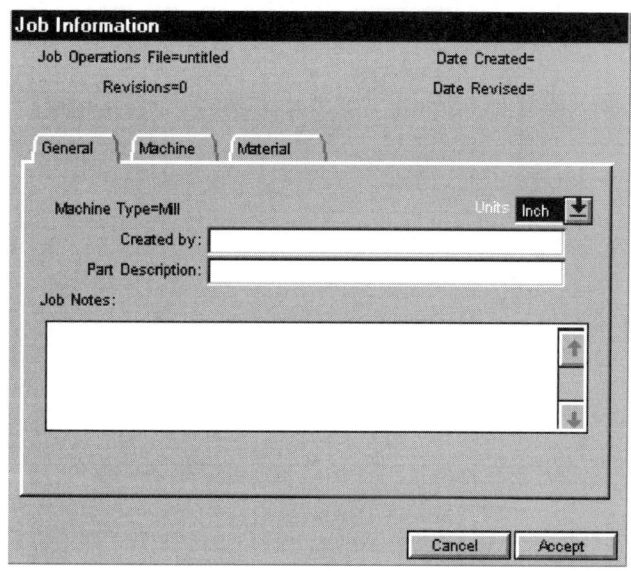

**Material** allows you to build a material library and specify material types in the event that you would need them to complete future projects.

Many of these input fields are for documentation purposes only. Input fields in the **Job Info** window that are left blank will not interfere with the completion of the remainder of the Process Planner nor will they interfere with the creation of the process model. You will, however, need to have a valid **SMF FILE** and a valid **TMP FILE**, found under **Machine**, before you can create machine code.

Once the input fields under **Job Info** are complete, close the **Job Info** window. You should now see a window with two tabs, the **Process Step List** and the **Tool List** as shown in Figure 2.5.

For this example, the **Process Step List** should be displayed. If it is not, simply choose the tab with your mouse and the **Process Step List** will

**FIGURE 2.5**
The **Process Step List** and the **Tool List**

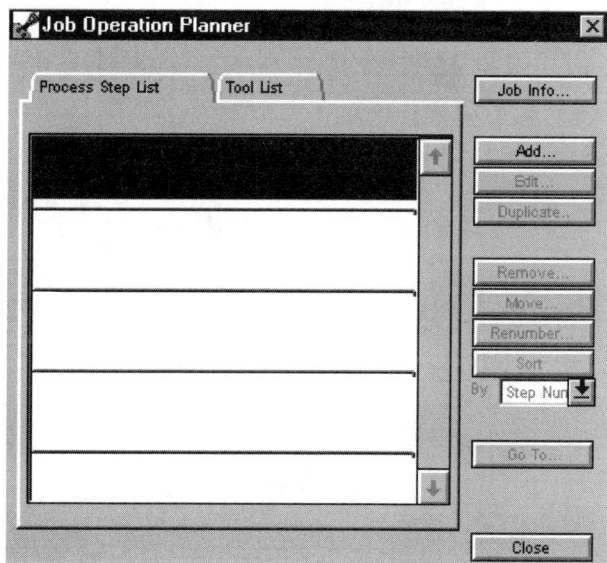

**FIGURE 2.6**
The **Add Process Step** window

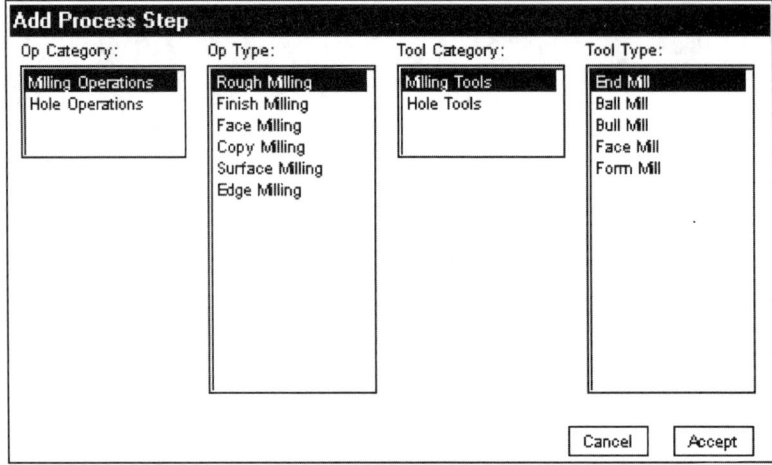

become the active window. To add steps to your list, choose the **Add** button. Upon choosing the **Add** button, the **Add Process Step** window should open. This window allows you to choose your cutting tools and define the type of cutting you will be doing. Figure 2.6 shows the **Add Process Step** window along with the highlighted selections you should make for this example.

Once the selections in the **Add Process Step** window are complete, choose the **Accept** button and the **Edit Process Step** window will open. It is here where the information concerning the tool and the operation of the tool is defined. Fill in the input fields of your **Edit Process Step** window to match those shown in Figure 2.7.

**FIGURE 2.7**
The **Edit Process Step** window

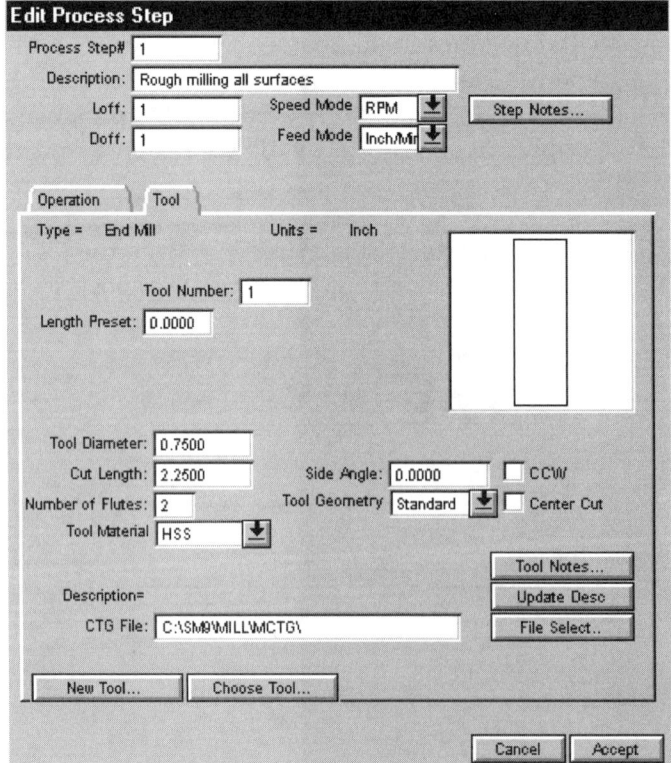

Many of the input fields of the **Edit Process Step** are self explanatory, and some will default to a standard value. A few of the fields, however, require an explanation of their values.

*Process Step #.*  This field is the number of the step that you are creating. If possible, it is best to create your steps in the order in which you will use them to machine your workpiece. However, there will be times when updates to your job file will require you to completely modify your job file. Methods are available to remove tools, move tools, and renumber tools. Those methods will be discussed later in this chapter.

*Description.*  This area allows for a brief description of the step that you will be using to construct the geometry.

*Loff/Doff.*  This is the number of the register in which the tool length offset and the diameter offset are stored. These values will normally default to the same number as that of the cutting tool. Special care should be taken to ensure that the Loff value is correct. Improper values could cause a crash at the machine tool. The actual values of the tool length offset (Loff) and cutter diameter compensation (Doff) will be entered in the corresponding register at the machine tool.

*Speed Mode.*  Either surface feet per minute (SFPM) or revolutions per minute (RPM) output to the code file is specified in this input field.

*Feed Mode.*  Either inches per minute (IPM) or inches per revolution (IPR) output to the code file is specified in this input field.

*Tool Number.*  This is the number of the cutting tool associated with the step you are currently defining. Tools will either be numbered in the order that you wish to use them, or in the order that you wish to arrange them in the machine tool carousel.

*Length Preset.*  This input field is to be used for prequalified tooling. If you are required to set the tool length offset value at the machine tool for each tool, this value should remain 0.000.

*Tool Diameter.*  This is the diameter of the cutting tool.

*Cut Length.*  This value represents the flute length of the cutting tool. This value should be as accurate as possible. When **Show Path** or **Show Cut** illustrates the cutting action of the cutting tool, the length of the graphics representation of the cutting tool depends on the value input in this field. If this value is accurate, it will allow you to determine if the flutes of the tool are long enough to properly machine your work.

*Number of Flutes.*  This field allows you to input the number of teeth on the cutting tool. This is necessary so that proper feed rates can be calculated once a chip load is input.

*Tool Material.*  This field will allow you to specify the cutting tool material. Choices are high-speed steel, carbide, cobalt, or ceramic.

*Side Angle.*  This field allows you to specify the angle of the cutting tool.

*Tool Geometry.*  This field allows you to specify the helix geometry of the cutting tool. Choices are Standard, Roughing, and Aluminum Cut.

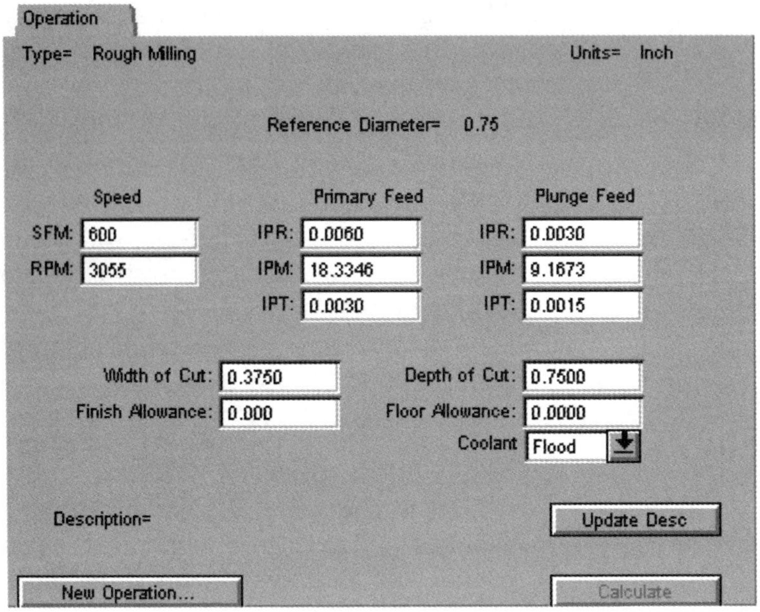

**FIGURE 2.8**
The input fields of the **Operation** page

Once the input fields of the **Tool** page are complete, choose the tab labeled **Operation**. This will open the **Operation** page and allow you to specify the speeds and feeds of the cutting tool. Fill in the input fields of the **Operation** page to match those shown in Figure 2.8.

*Speed.*   This allows you to input either the cutting speed for the material you are working with or the RPMs for the cutter. Either entry will automatically calculate the other.

*Primary Feed.*   This area allows you to input either inches per revolution, inches per minute, or inches per tooth feed rate for your cut. The primary feed rate is the feed generated in the "XY" plane. Again, either entry will automatically calculate the other.

*Secondary Feed.*   This is the feed in the "Z" axis. These values default as half the primary feed rate. If these default values are not appropriate for your application, a smaller input value is acceptable.

The default was accepted for the remaining input fields. The values may be adjusted for individual applications.

In this tutorial, as well as the remaining tutorials of this book, default values or other values that meet specific needs and applications are acceptable for the variety of input fields. However, all input fields must have valid input. Leaving input fields blank—especially those input fields in the job planner—could cause problems with processing the geometry of your model.

**Caution:** *Always fill in all input fields!*

Once the input fields are complete according to specifications, choose the **Accept** button in the lower right hand corner of the window. This will take you back to the original **Job Operation Planner** window and will show you the library of steps and tools that you are building. In order to add more steps to the library, simply choose the **Add** button.

**FIGURE 2.9**
The **Load Job File** dialog box

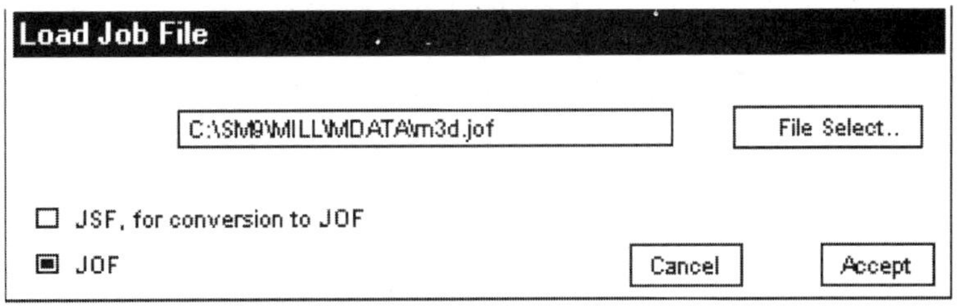

As was stated earlier, you may choose to read in a previously created job file. This may be one of the over fifty sample job files that come with the software, or it may be a file that you have previously created. Using this file may be a simple matter of choosing the tools you wish to use for this specific process model, or you may need to modify a job file to fit your needs.

To demonstrate the process of reading in a preexisting file and to demonstrate some additional features within the Job Operation Planner, choose **File** > **Load Job File**. (At this time, you will be prompted to save the job file on which you were previously working. Choose **No**, as you do not wish to save the previous example.) Fill in the **Load Job File** dialog box to match the one in Figure 2.9.

The "m3d.jof" job file should load and the thirty-three steps within this job file should now be seen in the database list.

You now have the capabilities of using these steps to construct the geometry of your process model. However, since this is a "preconstructed" job file, you need to take a few precautions. First you need to make sure that the tooling is numbered either in the order you wish to use it or by the number reflective of its location in your tool carousel. In addition, you need to check the speeds and feeds to make sure they are appropriate for your material.

For this example, assume that you wish to use Steps 1-9 and 21-30 of this job file to build your process model. You could simply choose those steps and build the geometry accordingly. However, in order to demonstrate a few more of the techniques of the job planner, modify this job file to more appropriately meet your needs.

The first step in modifying this job file is to delete all the steps you do not wish to use. To delete them, follow these steps:

1. Open the Job Planner by either choosing **File** > **Planner** or by choosing the Job Planner icon from the icon tool bar.

2. Choose the **Remove** button from the right of the window. The **Remove** window will open and will show all steps that are eligible to be removed. Steps that are ineligible are those which have geometry associated with them. In addition, you cannot remove the step that is considered active in the **Insert** control panel.

3. Inside the **Remove** window, choose each step that you wish to remove. These will be steps 10-20 and steps 31-33. Notice that once you choose the step, an asterisk appears on the left side of the step description to "tag" the step for removal. Once all the asterisks are in place, choose the **Remove** button and the "tagged" steps will be removed.

4. Choose the **Close** button and the **Remove** window will close.

The next step to modifying your job plan is to remove the tools that were associated with the steps that were just removed. To remove those tools, follow these steps:

1. Choose the **Tool List** tab inside the **Job Planner** window.
2. Choose the **Remove** button. Only those tools that are not associated with a step will be listed.
3. Choose the **Remove All** button.

At this time, it is a good idea to save the updated version of your job planner. To save, follow these steps:

1. Choose **File**.
2. Choose **Save Job File.**
3. Fill in your **Save Job File** dialog box to match the one in Figure 2.10.
4. Choose the **Accept** button and the updated file will be saved.

The next modification is the addition of two new steps. A face mill operation and a drill operation are to be added to your database list. To add them, follow these steps:

1. Open the job planner by one of the two methods previously explained.
2. If the **Process Step List** window is not already active, choose the **Process Step List** tab to activate it.
3. Highlight the very last step, step 30.
4. Choose the **Add** button.
5. Create a step that contains the information in the following table:

**TABLE 2.1**

| Step # | Tool # | Type | Diameter | Speed | Feed |
|---|---|---|---|---|---|
| 31 | 31 | Twist Drill | 3/8″ | 90 SFPM | 12 IPM |

**FIGURE 2.10**
The Save Job File

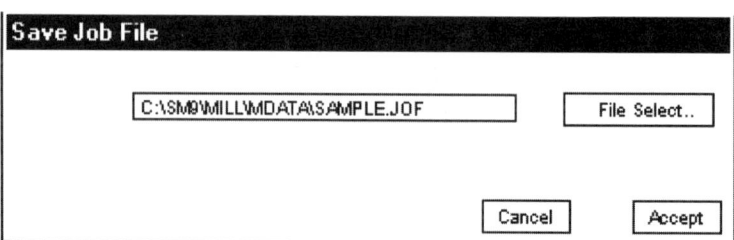

Chapter 2   SmartCAM Tutorial 1

6. To complete the addition of the twist drill, choose the **Accept** button.
7. Once the addition of the twist drill is complete, highlight it with the mouse and choose the add button. You will now add a face milling step that contains the information in Table 2.2:
8. To complete the addition of the face mill, choose the **Accept** button.

At this point, your job planner window should list the original steps 1-9 and 21-30 along with the added steps 31 and 32.

In this text's tutorials that require the use of a face mill, it is used first in sequence. For the sake of simplicity, it is a good idea to order the steps in the sequence in which you use them. To further modify this job plan to meet your needs, move the face mill step to the beginning of the list. To move it to the beginning, follow these steps:

1. Highlight step 32 with the mouse.
2. Choose the **Move** button.
3. When the **Move Process Step** dialog box opens, select the "To beginning" option.
4. Choose the **Accept** button, and step 32 will be repositioned at the beginning of the list.

Next, you will need to renumber the steps. To renumber, follow these steps:

1. Choose the **Renumber** button.
2. When the **Renumber Process Step** dialog box opens, input a beginning number of "1" and an increment of "1."
3. Choose the **Accept** button.

At this time, the steps in your list should be renumbered in sequential order. However, if you will look closely, you'll see that the tool numbers were not renumbered when the steps were renumbered. Again, two options exist for numbering the tools. The tools can either be numbered to agree with the step they are associated with or they can be numbered according to their position in the machine tool carousel. For this example, simply number the tools so they agree with the step with which they are associated.

The process of renumbering the tools is a little more difficult than the process of renumbering the steps. Logic would tell you to simply pick the first step, choose the edit button, and change the tool number to agree with the step number. However, since tool 1 is already assigned to step 2, you will get the error message shown in Figure 2.11.

**TABLE 2.2**

| Step # | Tool # | Type | Diameter | Speed | Feed |
|---|---|---|---|---|---|
| 32 | 32 | Face Mill | 3.0" | 763 RPMs | 30 IPM |

**FIGURE 2.11**
Tool numbering error message

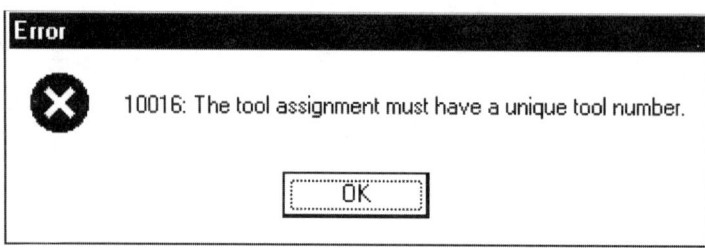

Normally, the best way to renumber the tools is to start with the last tool in the list—which should be the step with the largest number—and edit the tool number. In this example, however, this technique will present a problem since step 11 already has tool 21 assigned to it. If you attempt to change the tool number input field to 21, you will get the error message that was shown in Figure 2.11.

To correctly renumber the tools in this example, follow these steps:

1. Highlight step 20.
2. Choose the **Edit** button.
3. Change the values in the **Tool Number** input field, the **Loff** input field, and the **Doff** input field to agree with the step number.
4. Choose the **Accept** button.

The Job Operation Planner window will reveal the updates. Tool 20 should now be associated with step 20. Continue with step 19 down to step 1. In each situation, edit the **Tool Number** and the **Loff** and **Doff** values to match the appropriate step number. At this time, you may also want to edit any speeds and feeds which you feel are incorrect for your process model.

Once the updates for step 20 through step 1 are complete, return to step 21 and update that step in the same manner as was done in the previous steps. Since step 11 now has tool 11 associated with it (instead of tool 21), there will be no problem when you attempt to update step 21.

At this time, the Job Operation Planner window should list steps 1 through 21. Each step should have a tool associated with it that has the same number.

In order to have a full understanding of all the functions of the Job Operations Planner, there are two additional features that need to be discussed. The first is the **Duplicate** button to the right of the Job Operation Planner window.

To use the Duplicate button, follow these steps:

1. Highlight the step which you wish to duplicate.
2. Choose the **Duplicate** button.

The Edit Process Step window will open as normal. Although the window will show a new process step number, all other fields will be an exact duplicate of the highlighted step. You may choose to edit this new process step or you may wish to use the original values.

Chapter 2   SmartCAM Tutorial 1                                              **31**

The **Choose Tool** button is the last feature that will be discussed in this tutorial. It is very similar in function to the duplication feature that was just discussed. However, the **Choose Tool** button will only allow the duplication of the information associated with the tool, not the operation of the tool. This is the true concept of the principle of the step in the respect that you want to use the same cutting tool, yet operate that tool under two distinctly different conditions.

To demonstrate the use of the **Choose Tool** button, follow these steps:

1. Highlight the last step in the list, step 21.
2. Choose the **Add** button.
3. When the **Add Process Step** window opens, make the appropriate choices for a rough milling operation using an endmill.
4. When the **Edit Process Step** window opens, change the value in the **Process Step #** input field to 22.
5. At the bottom of the **Tool** window, which is open within the **Edit Process Step** window, choose the button that is labeled **Choose Tool.** The **Choose Tool** window will open and will list all the tools that have been previously defined as endmills.
6. Choose tool 17.
7. Pick the **Use** button at the bottom of the window.

   The physical properties associated with this tool will be entered into your current process step.
8. Continue by choosing the **Operation** tab.
9. Input a cutting speed of 600 SFPM and a primary feed of 25 IPM.
10. Accept the default for all other fields.
11. Choose the **Accept** button.

At this time, the Job Operation Planner window should list 22 steps. Steps 1 through 21 should list tool numbers that are the same as the step numbers with which they are associated. Step 22 should list tool 17 as its tool.

12. Choose **Close** to exit the Job Operation Planner window.

# CHAPTER 3

# SmartCAM Tutorial 2

## Slotted Plate

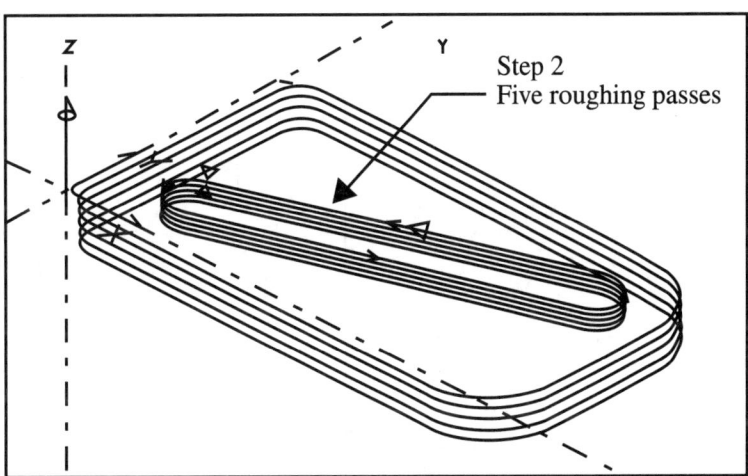

# Chapter 3  SmartCAM Tutorial 2

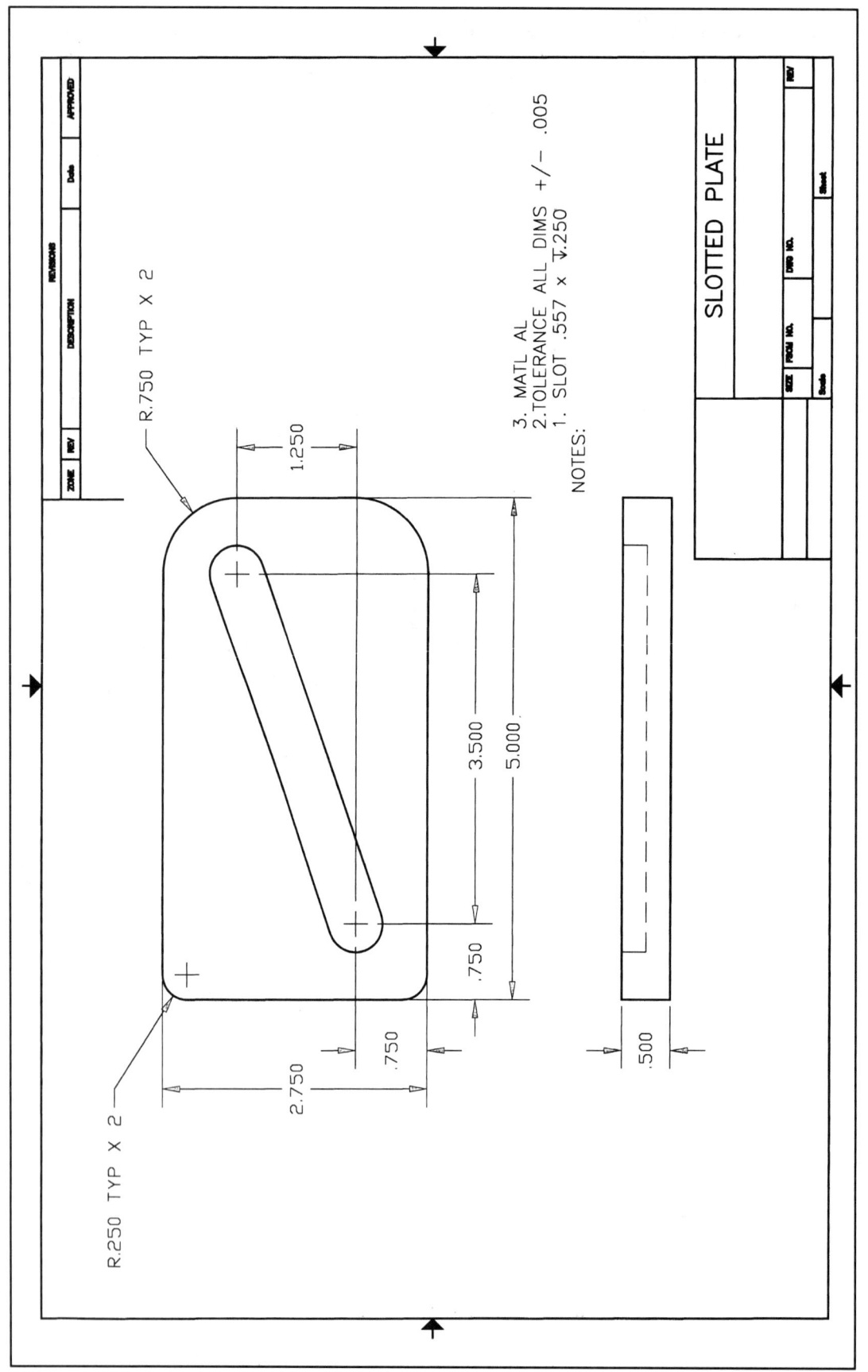

The blueprint for the Slotted Plate Tutorial

# Chapter 3  SmartCAM Tutorial 2

Upon completion of this chapter, you should be able to:

- Correctly build a job file from instructions.
- Construct basic geometry utilizing the snap functions.
- Edit geometry to add various size blends.
- Group geometry using both Step and Profile methods.
- Create multiple profiles using the Transform function.
- Create geometry which is tangent to existing geometry.
- Trim geometry to create profiles.
- Correctly sequence steps in the database.

# SLOTTED PLATE

This tutorial is meant to instruct you in the basic processes of geometry construction. Its purpose is to build foundational skills to assist you in future tutorials. All processes will be kept as simple as possible so that you will understand the steps required to correctly create a process model. Operations that are more advanced in nature will occur in future tutorials.

In building a process model, you are required to have an active job file that lists all the tools you wish to use before you can construct "toolpath" geometry. Although you were introduced to the Job Operation Planner in the previous tutorial, this tutorial will also step you through the Job Operation Planner in order to reinforce the previous lesson.

In order to build the job file that is required for this tutorial, open the Job Operation Planner. Either choose the job planner icon from the icon bar or choose **File > Planner**. Upon opening the job planner, choose the **Job Info** button in the upper right corner of the window.

Figure 3.1 shows the open Job Operation Planner along with the location of the Job Info button.

The job information area of the Job Operation Planner allows you to specify information about the process model in three areas: the General page, the Machine page, and the Material page.

The **General** page allows you to document general information such as the process model's unit of measure, the creator of the file, and various notes concerning your process model. Enter the information shown in Figure 3.2 into your **General** page.

As you may recall from the previous tutorial, the information in the **General** page is optional and is primarily used for documentation purposes.

The **Machine** page, opened by selecting the **Machine** tab with your mouse, allows you to specify the correct Machine Define (SMF) file and the correct Template (TMP) file. These two files are responsible for generating the correct "G" code for your specific machine tool. This text does not include information concerning the template file and the machine file. However, information concerning these two files may be found under the **Help** menu of SmartCAM.

If you are intending to generate machine code, you must enter a valid file name into these two input fields. Failure to do so will result in an error when you attempt to generate "G" code.

**36** Chapter 3 SmartCAM Tutorial 2

**FIGURE 3.1**
The Job Info button

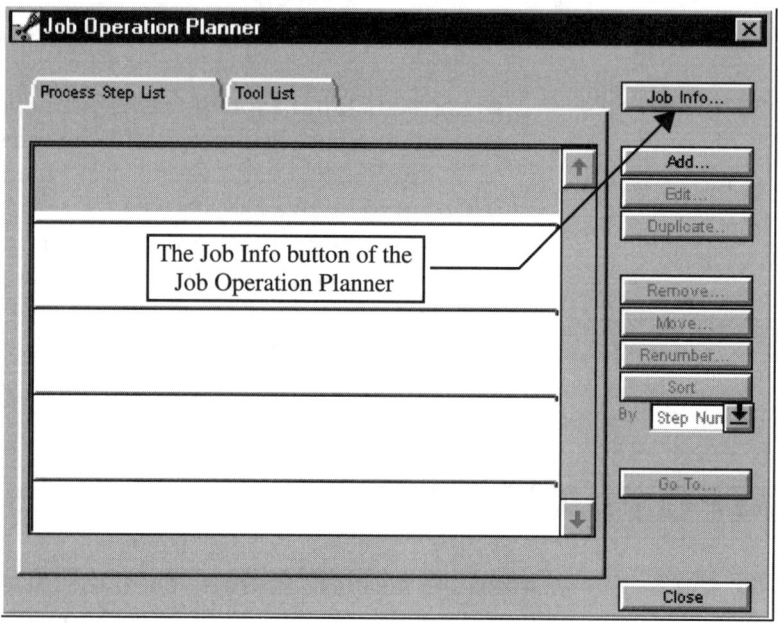

**FIGURE 3.2**
The General page

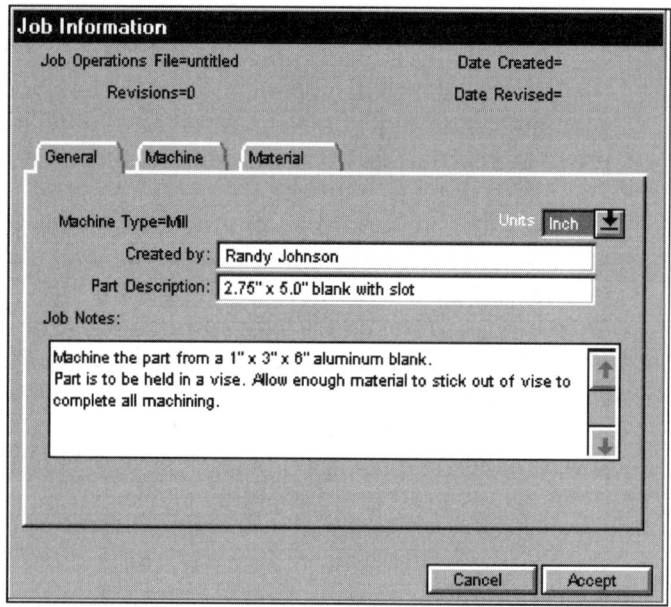

To input the correct SMF File, follow these steps:

1. Highlight the word "SMF File" by selecting it with your mouse as shown in Figure 3.3.
2. Pick the **File Select** button with your mouse.
3. Select the appropriate file from the MSMF subdirectory. (If the files have not been moved from their original location, they will be found in the C:\SM9\MILL\MSMF subdirectory.)
4. Choose **Accept**.

Chapter 3  SmartCAM Tutorial 2    37

**FIGURE 3.3**
The SMF File input field of the Machine page

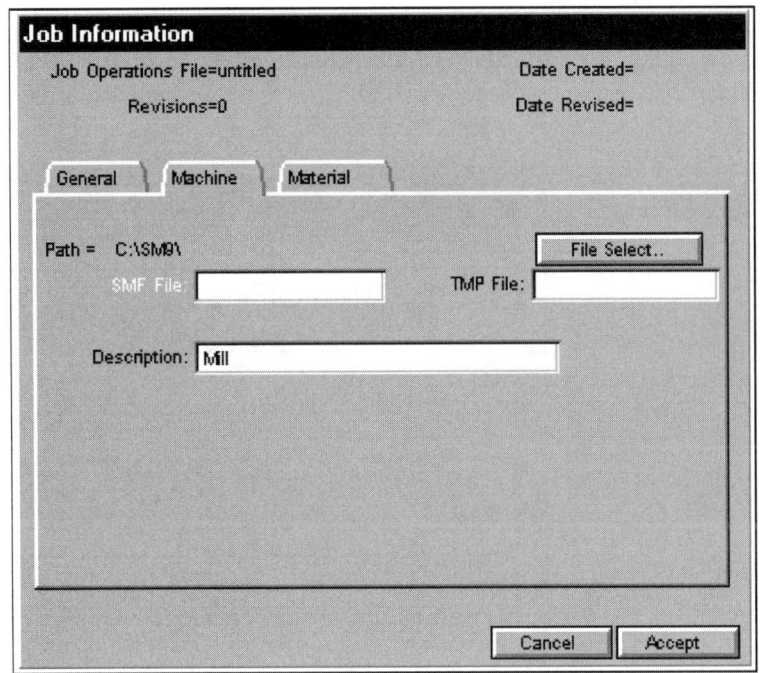

The correct SMF file should appear in the SMF File input field of the **Machine** page.

5. Next, highlight the TMP input field by choosing the word "TMP File" with your mouse.
6. Pick the **File Select** button with your mouse.
7. Select the appropriate file from the MSMF subdirectory, as was done before.
8. Choose **Accept**.

The correct TMP file should now appear in the TMP File input field of the **Machine** page.

9. When both input fields are complete, choose the **Accept** button at the bottom right of the window.

(The Material page will not be covered in the context of this manual. Information concerning the Material page can be found under the Help menu of SmartCAM.)

The next operation is to input the steps into the Job Operation Planner. The first step you will use to construct your geometry will consist of a 1/2" diameter 2 flute endmill. You will use this tool to first rough the perimeter of the part and then rough the slot.

At this time, the Job Operation Planner should be open as shown in Figure 3.4.

To input the steps into the Job Operation Planner, follow these steps:

1. Choose the **Add** button to add steps to the Process Step List. Upon choosing the **Add** button, the **Add Process Step** dialog box will

**FIGURE 3.4**
The Add button of the Job Operation Planner

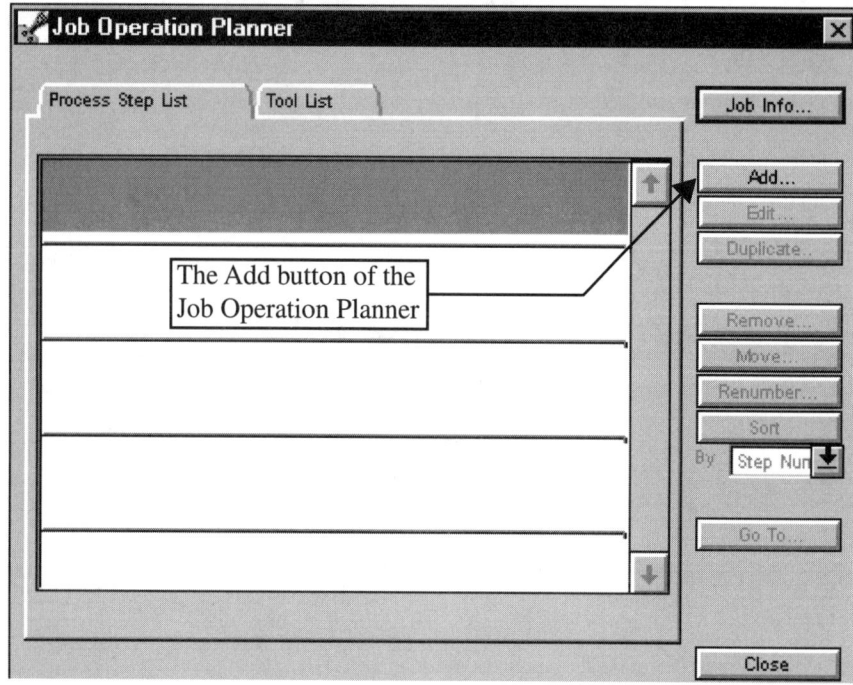

open. Figure 3.5 shows the **Add Process Step** dialog box along with the selections you should make for this example.

2. Once the selections in the **Add Process Step** dialog box are complete, choose the **Accept** button and the **Edit Process Step** dialog box will open. Fill in the input fields of your **Edit Process Step** dialog box to match those shown in Figure 3.6.

3. Once the Tool page of the **Edit Process Step** dialog box is completed, proceed to fill in the **Operation** page as shown in Figure 3.7. (To open the **Operation** page, simply choose the tab with your mouse.)

**FIGURE 3.5**
The Add Process Step dialog box

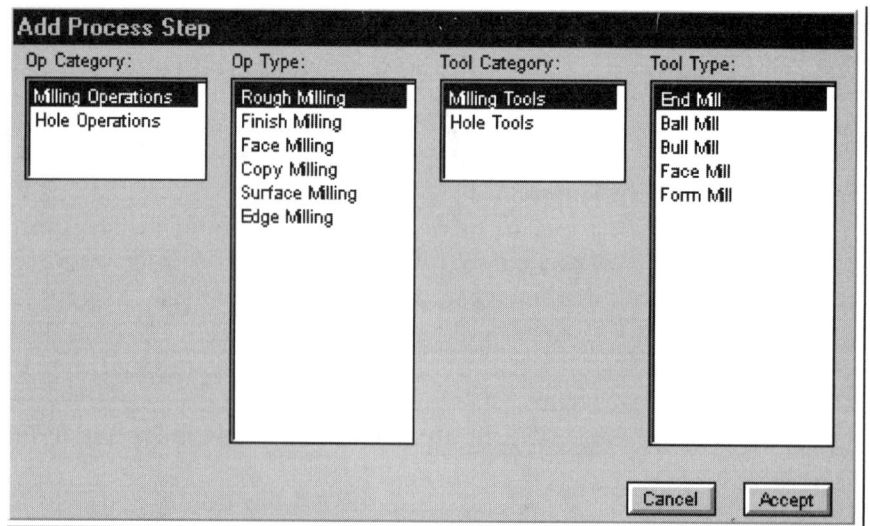

# Chapter 3  SmartCAM Tutorial 2

**FIGURE 3.6**
The Edit Process Step dialog box with open Tool page

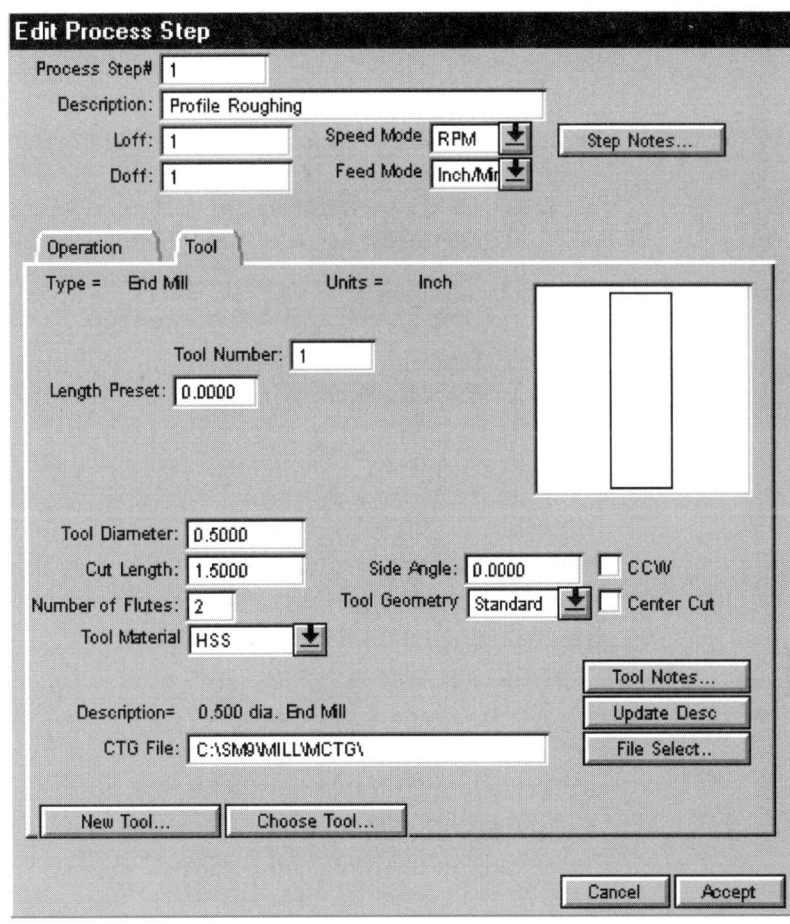

**FIGURE 3.7**
The Operation page of the Edit Process Step window

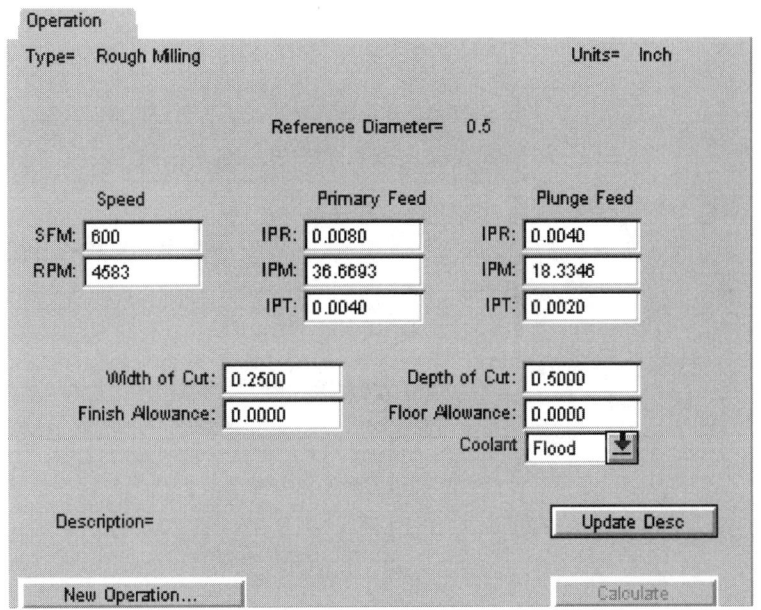

4. Once the **Operation** page is complete, choose the **Accept** button from the lower right corner of the **Edit Process Step** dialog box.

At this time the Job Operation Planner will show step 1 listed in the library. Your Job Operation Planner should match that shown in Figure 3.8.

5. Continue building the job file by adding the second step. Choose the **Add** button of the Job Operation Planner.
6. When the Add Process Step dialog box opens, again make the selections so your **Add Process Step** dialog box matches that shown in Figure 3.5.
7. Choose **Accept**.

The second step of your process model will use the same tool as did step 1. Recall from the previous tutorial that assigning the same tool to different steps will give you independent control of the speeds and feeds of the tool when used in two different applications. Tool 1 will be used to rough the peripheral edges of the part as well as the slot. Since machining the slot requires more tool engagement, you need to reduce the feed rate.

**CAUTION!** *If you attempt to type in all information in the **Edit Process Step** dialog box, as was done for step 1, you will receive the error message shown in Figure 3.9.*

When you type in information in each input field, the SmartCAM system actually attempts to create another tool. Since tool 1 already exists, the previous error message will be displayed.

To correctly assign tool 1 to step 2, follow these steps:

1. Enter 2 into the **Process Step #** input field.

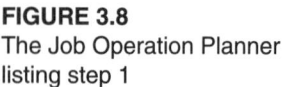
**FIGURE 3.8**
The Job Operation Planner listing step 1

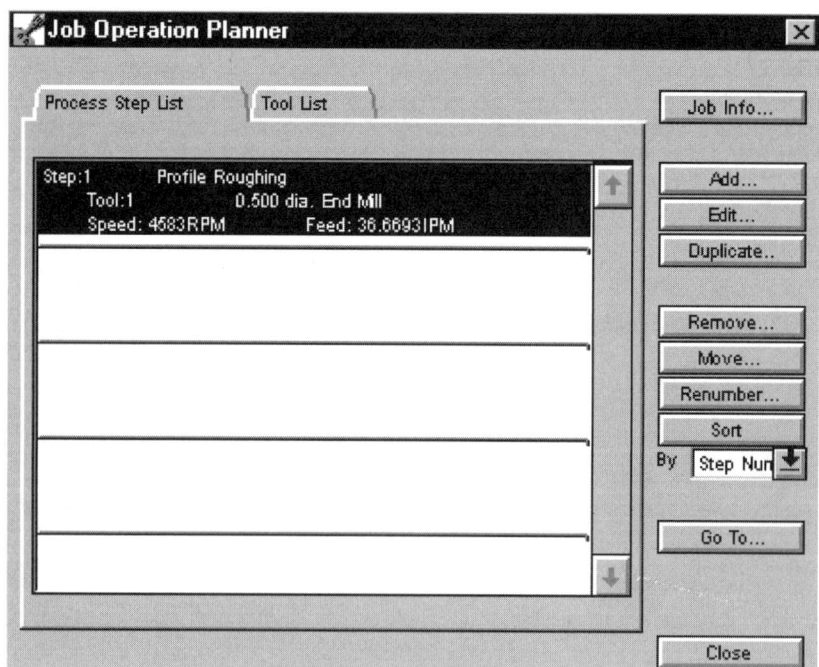

Chapter 3  SmartCAM Tutorial 2    **41**

**FIGURE 3.9**
Edit Process Step dialog box error message

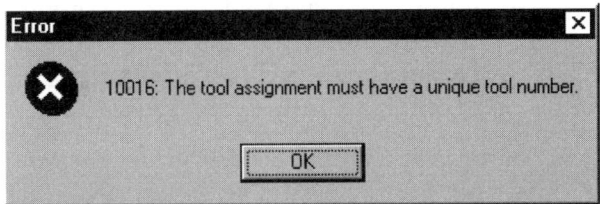

2. Enter "Slot Roughing" into the **Description** input field.
3. Select the **Choose Tool** button at the bottom of the Tool page as shown in Figure 3.10.

Upon selecting the **Choose Tool** button, the **Choose Tool** dialog box opens as shown in Figure 3.11.

**FIGURE 3.10**
The **Choose Tool** button of the Tool page

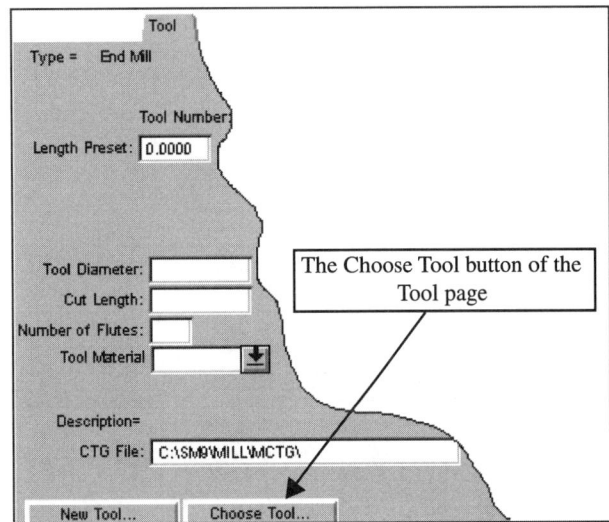

**FIGURE 3.11**
The **Choose Tool** dialog box and the Use button

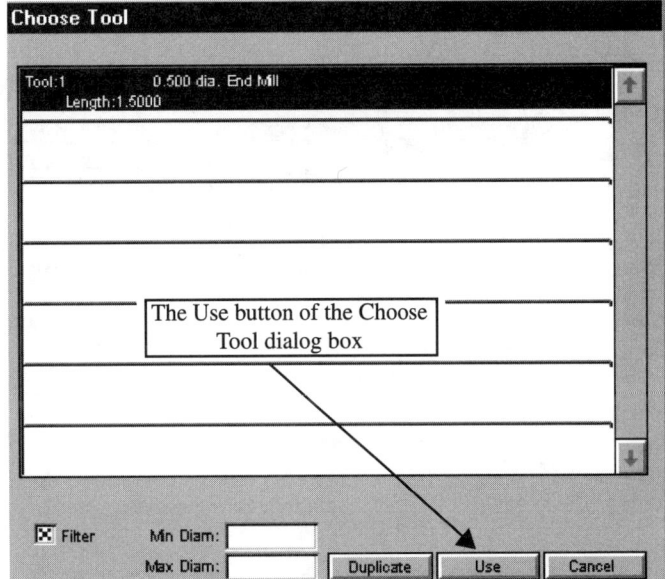

4. Simply highlight tool 1 by selecting it with your mouse and select the **Use** button as shown in Figure 3.11. The input fields of the **Tool** page will be filled out with the properties of tool 1.
5. Continue adding tool 1 to step 2 by completing the **Operation** page as shown in Figure 3.12.
6. When the input fields of the Operation page are complete, choose the **Accept** button at the bottom of the Job Operation Planner.

To create step 3, the finishing tool, follow these steps:

1. Choose the **Add** button of the Job Operation Planner.
2. When the **Add Process Step** dialog box opens, make the appropriate selections as shown in Figure 3.13.

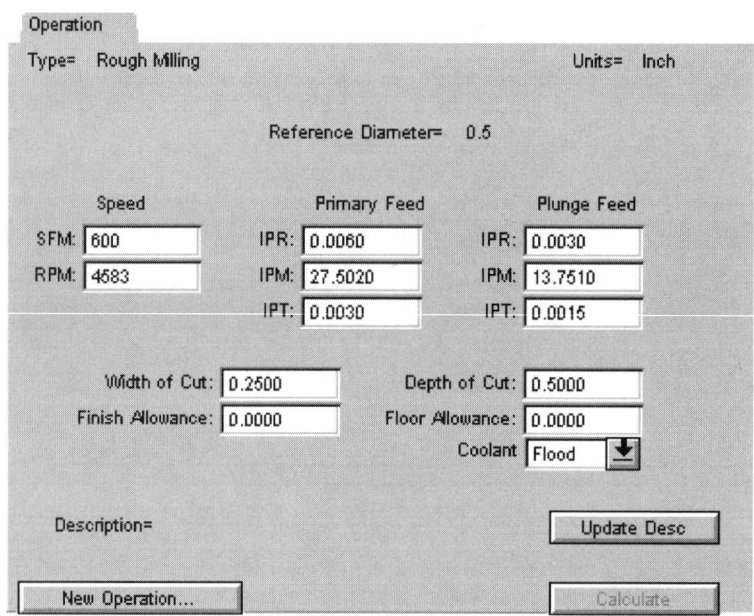

**FIGURE 3.12**
The **Operation** page for step 2

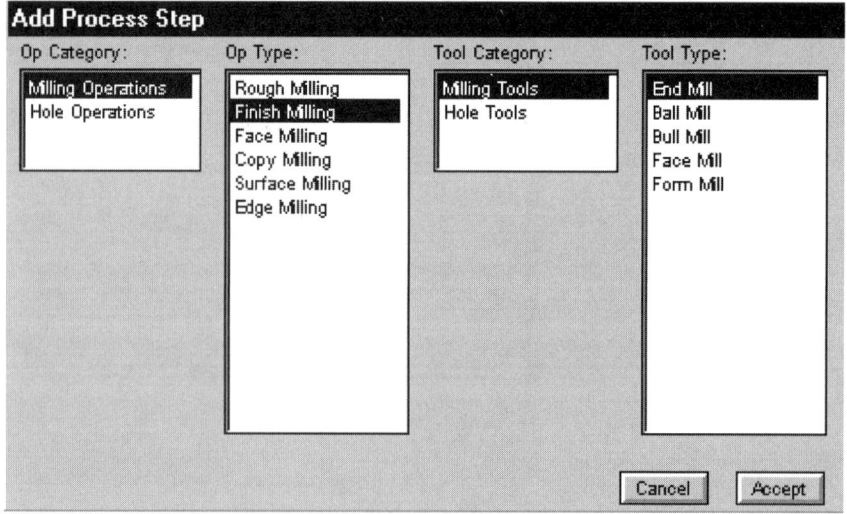

**FIGURE 3.13**
The **Add Process Step** dialog box for step 3

# Chapter 3  SmartCAM Tutorial 2

3. Choose the **Accept** button in the lower right corner.
4. Continue creating step 3 by filling in the input fields of your **Edit Process Step** dialog box to match Figure 3.14.
5. Continue creating step 3 by choosing the Operation tab of the **Edit Process Step** dialog box and fill in the input fields as shown in Figure 3.15.

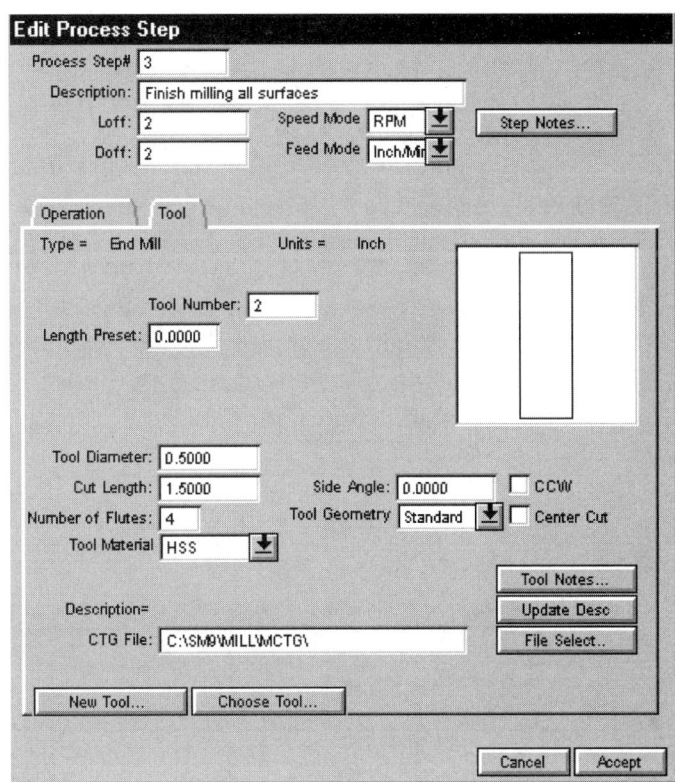

**FIGURE 3.14**
The **Edit Process Step** dialog box for step 3

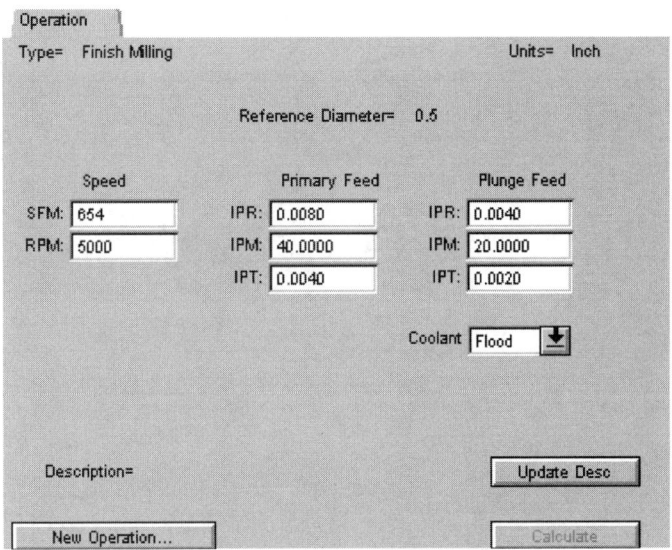

**FIGURE 3.15**
The Operation page for step 3

**44** Chapter 3 SmartCAM Tutorial 2

6. To complete the Job Operation Planner for the Slotted Plate, choose the **Accept** button at the bottom of the Job Operation Planner.

The library that you have just completed should match Figure 3.16.

# Creating the Finish Profile

As with all SmartCAM process models, the construction begins by first creating the finish profiles. This is done because the blueprints from which you work show finish geometry. You construct the finish geometry first, then resequence the database to position the roughing cuts in front of the finish cuts.

To begin constructing the finish profile, draw a 2.75" by 5.00" rectangle, following these steps:

1. Choose the **Insert** toolbox from the workbench.
2. The **After** selection and the **Element Seq** tool should be chosen as the default since the only other element in the database is the origin.
3. You will need to machine the finish profile, therefore, the **With Step** selection should be "on." The filled-in circle indicates the on condition.

4. Fill in the input fields of your **Insert** control panel to match those shown in Figure 3.17.

**FIGURE 3.16**
The completed library of steps

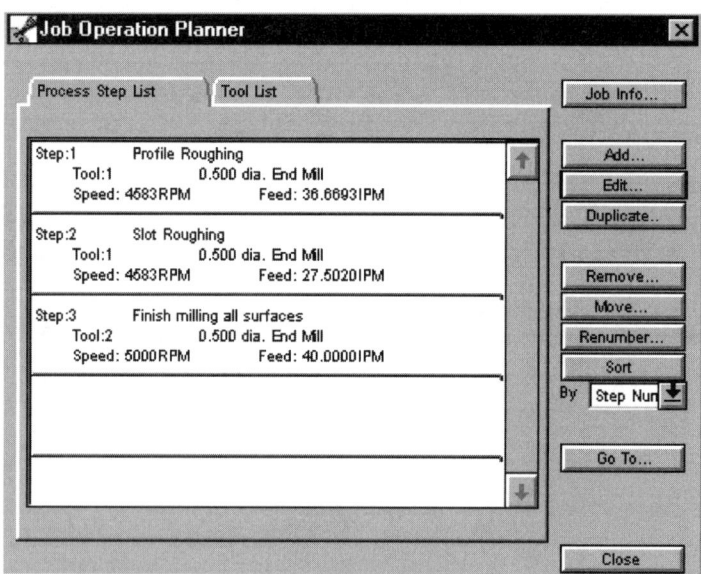

**FIGURE 3.17**
The **Insert** control panel

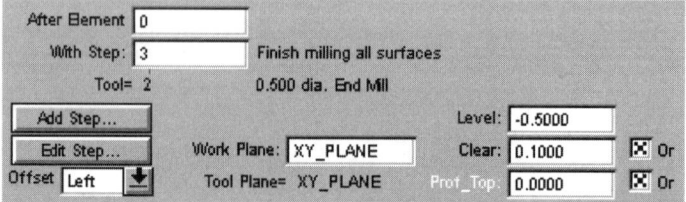

Chapter 3    SmartCAM Tutorial 2

**After Element** is set to 0 because there are no other elements in the database at this time.

**With Step** is set to 3, the finish step. Remember, construct the finish geometry first.

**Offset** is set to Left. Your geometry will be constructed in a clockwise manner. A left offset will allow you to climb mill.

**Work Plane** and **Tool Plane** will default to the "XY" Plane.

**Level** is set to -.500", the depth of the finish cut.

**Clear** is set to .100". This value is the point at which the cutting tool will retract before making any rapid traverse moves. Any appropriate value will work.

**Prof Top** is the "Z" level of the surface of the workpiece. In this example, 0.000" is the top of the work.

5. Choose the **Geometry** toolbox from the workbench. Remember, if a specific toolbox is not on the workbench, it can be found under the menu selections.
6. Choose the **Line** modeling tool from the tool list.
7. Fill in the **Line** control panel to match the one shown in Figure 3.18. Highlight each individual input field by selecting it with your mouse. This will make that input field "active" and ready for input. Enter the appropriate values according to Figure 3.18 by typing the value and then pressing the enter key.

If you will look closely at the "Y" input field, you will see a small asterisk (*) slightly to the left of the "Y." This is referred to as a "trigger" in SmartCAM. The trigger lets you know that upon completion of that particular input field, SmartCAM will have enough information to complete the task.

At this time, you should see a vertical line in your graphics work area.

To create the second line, you will work with the snap commands as well as the filters.

1. If it is not already, turn on the **Automatic** mode by selecting both the Free Coordinate Mode and the Snap Mode buttons from the Snap Icon area. Remember, these icons are on when they appear to be pressed in.
2. Select the endpoint icon.
3. Select the Edit Filter icon.
4. When the Edit Filter dialog box opens, select the **All** button. This will allow the selection of all element types.

**FIGURE 3.18**
The **Line** control panel

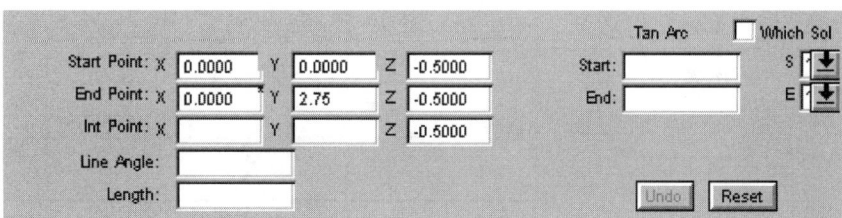

5. Select the **Accept** button.
6. Again, from the **Line** control panel, highlight the word "Start Point" by selecting it with your mouse.
7. Move your mouse to the top portion of the vertical line. When the cursor switches from the larger crosshairs to the smaller crosshairs, click the left mouse button. This will "snap" to the endpoint of the line and will input all the values of that point into the **Start Point** input fields.
8. Highlight the "X" input field of the **End Point**. Input 5.00 taken from the blueprint.
9. Highlight the "Y" input field of the **End Point** (notice the asterisk). Input 2.75 from the blueprint.

Your screen should now show two lines as shown in Figure 3.19.

Continue the process of creating geometry until you have a rectangle as shown in Figure 3.20.

Continue with the finish profile by adding the two 1/4" radii. To do so, follow these steps:

**FIGURE 3.19**
Elements 1 and 2

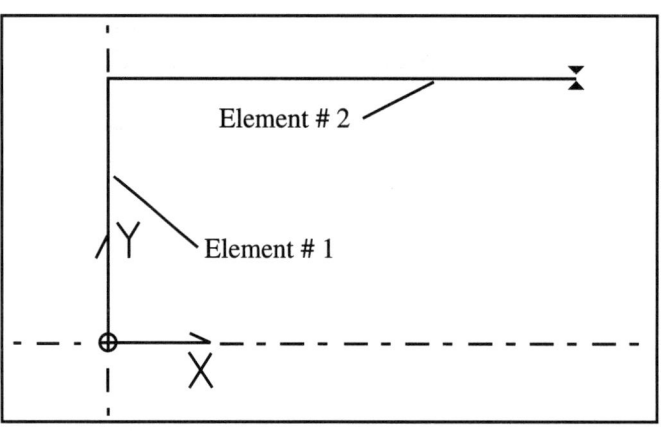

**FIGURE 3.20**
The completed rectangle

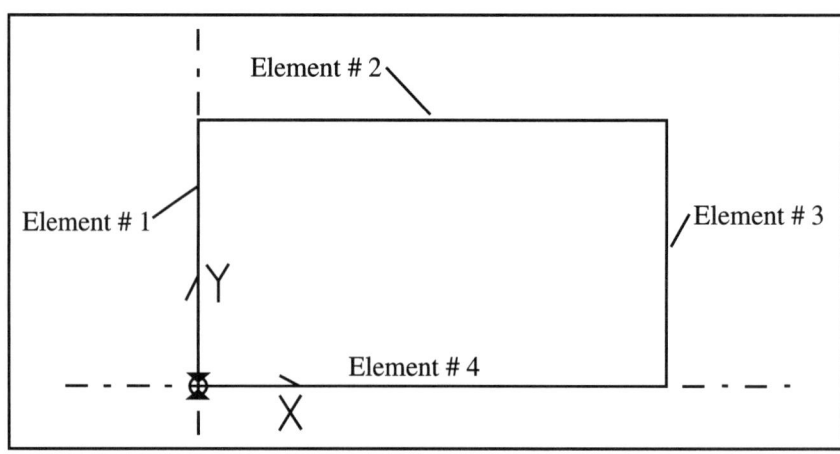

1. Choose the **Geo Edit** toolbox from the workbench.
2. Choose the **Blend** tool from the tool list.
3. The first input field that must be filled in is the **Inside Radius** input field. It should default to .250. If it does not, input .250 into this field.
4. Highlight the **Select 1st Element** input field. Type the number 1 into this input field.
5. Highlight the **Select 2nd Element** input field. Enter 2 into this input field.

The blended radius should match Figure 3.21.

6. Again, referring to the **Blend** control panel, highlight **Select 1st Element**. Enter 5 into this input field.
7. Highlight **Select 2nd Element**. Enter 1 into this input field.

Your process model should now match Figure 3.22.

The next process will input the two 3/4" radii in exactly the same way as the 1/4" radii were placed in the process model.

1. Again, in the **Blend** control panel, input .750 in the **Inside Radius** input field.

**FIGURE 3.21**
The 1/4" blend between elements 1 and 2

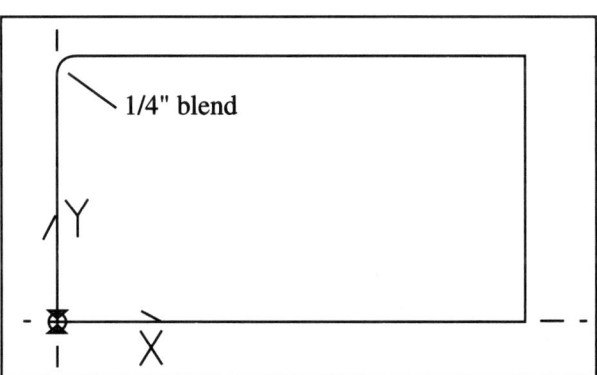

**FIGURE 3.22**
The two 1/4" blends in place

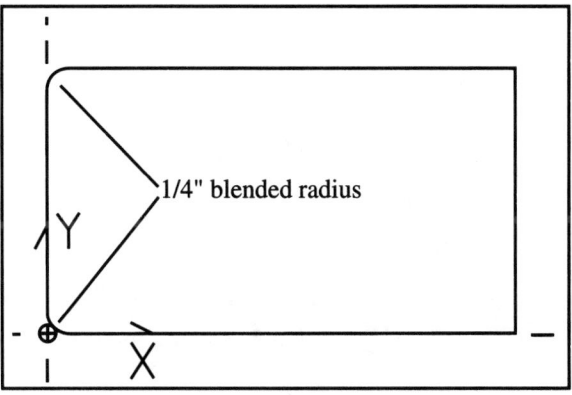

2. Highlight **Select 1st Element**. Select the upper horizontal line.
3. Highlight **Select 2nd Element**. Select the right vertical line.
4. Highlight **Select 1st Element**. Select the right vertical line.
5. Highlight **Select 1st Element**. Select the lower horizontal line.

At this time, your process model should match Figure 3.23.

## Constructing the Slot Profile

The next step in constructing the finish geometry is to construct the slot. This will be done by first creating the two radii as full arcs, constructing two lines tangent to the two arcs, and then trimming the two arcs to the tangent point.

To do so, follow these steps:

1. Choose **Insert** from the workbench.
2. Fill in the input fields of your **Insert** control panel to match those in Figure 3.24.

(Notice the **Level** changes to reflect the "Z" level of the slot.)

3. Choose **Geometry** from the workbench.
4. Choose **Arc** from the tool list.
5. Fill in the input fields of the **Arc** control panel to match those in Figure 3.25.

**FIGURE 3.23**
The complete exterior finish profile

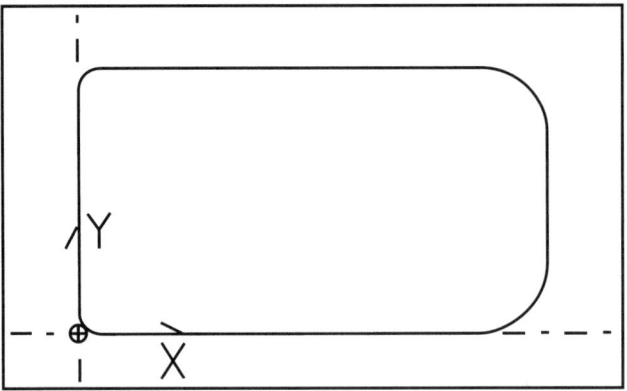

**FIGURE 3.24**
The **Insert** control panel with settings for the slot

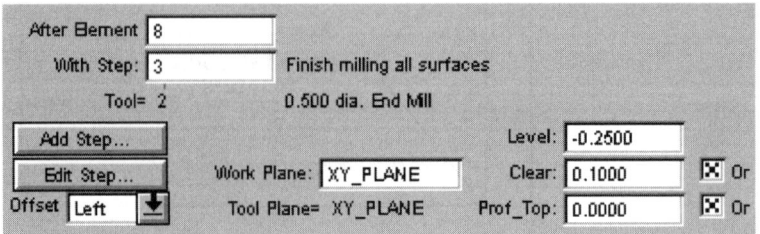

**FIGURE 3.25**
The **Arc** control panel

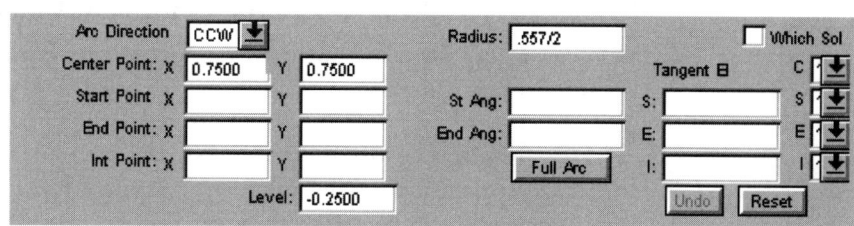

**Arc Direction** is set to counterclockwise (CCW) so that as you mill the inside of the arc, you are climb milling.

**Center Point** contains values that are entered directly from the blueprint.

**Level** This input field will contain a value that is carried over from the **Insert** control panel.

**Radius** contains the mathematical formula .557/2. When you enter this value, the CAM system will carry out this mathematical operation. The result is .2785. Figure 3.24 showed the mathematical operation that illustrates this function of SmartCAM. SmartCAM will carry out all mathematical functions when entered into the input fields correctly.

At the completion of these input fields, SmartCAM will show a potential solution of the arc. (Notice the dashed lines representing the arc.)

6. The last step is to select the **Full Arc** button.

An arc should now be shown in the lower left portion of the process model.
To create the next arc, simply input .75+3.5 for the "X" input field and .75+1.25 for the "Y" input field.
At this time, your process model should match Figure 3.26.
Continue with the construction of the slot by inserting two lines that are tangent to the two arcs.

1. Choose **Insert** from the workbench.
2. Select **After** and **Element Seq** from the tool list.

**FIGURE 3.26**
The completed arcs

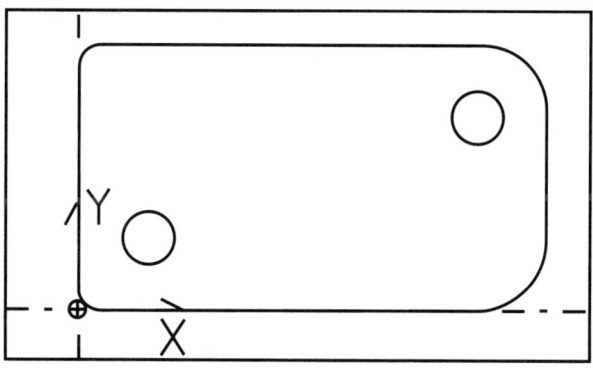

3. Highlight **After Element** from within the **Insert** control panel.
4. Choose the bottom arc. Accept the default values for all other input fields.

The purpose of the previous four steps is to resequence the database list. It's important to understand the purpose of resequencing the database as you just did. The line you are about to create will be machined after the arc. Therefore, the line must come after the arc in the database. The elements will be machined according to their order in the database. If the elements are out of order in the database, the result will be excessive rapid traverse moves. Elements that are severely out of sequence could crash the machine tool.

5. Choose **Geometry** from the workbench.
6. Choose **Line** from the tool list.

To create this line, you will work with the Tan Arc (Tangent to an Arc) function in the **Line** control panel as shown in Figure 3.27.

7. Choose the **Start** input field of the Tan Arc function.
8. Choose the lower arc according to Figure 3.28.
9. Check to verify the **End** input field is highlighted. If it is not, select it with the mouse to highlight it.
10. Select the upper arc according to Figure 3.29.

At this time, your process model should match Figure 3.30.

To insert the next line, you must resequence the database as was done earlier.

**FIGURE 3.27**
The Tan Arc function

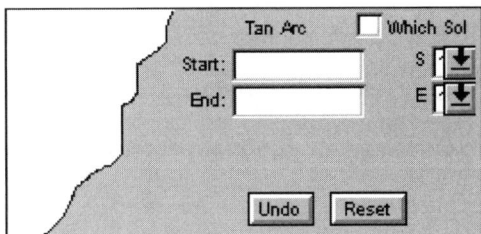

**FIGURE 3.28**
The Tan Arc Start location

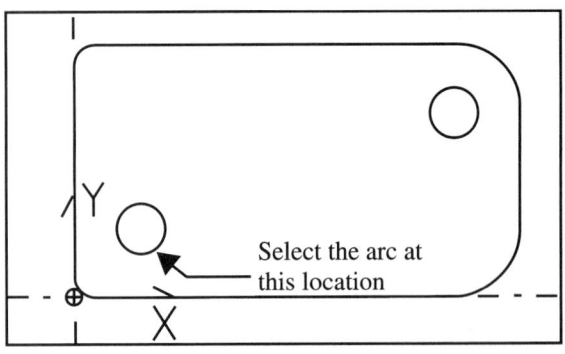

**FIGURE 3.29**
The Tan Arc End location

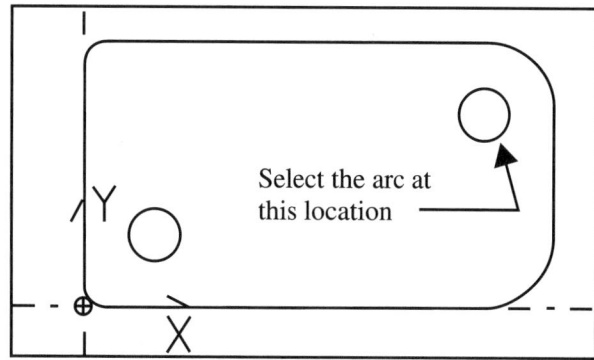

**FIGURE 3.30**
The line constructed tangent to the arcs

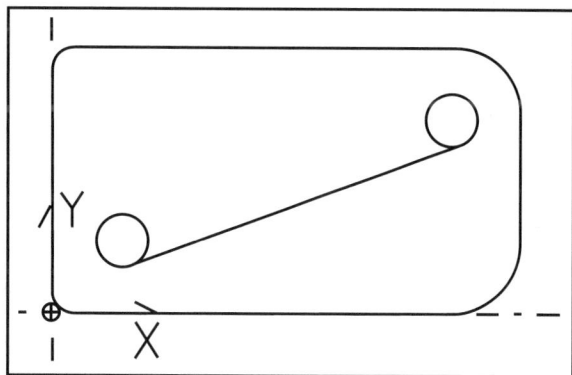

11. Choose **Insert** from the workbench.
12. Select **After** and **Element Seq** from the tool list.
13. Highlight **After Element** from within the **Insert** control panel.
14. Choose the upper arc from the graphics work area.
15. Choose **Geometry** from the workbench.
16. Choose **Line** from the tool list.
17. From within the **Line** control panel, choose the **Start** input field, as was done in the previous example.
18. Choose the upper arc as shown in Figure 3.31.

**FIGURE 3.31**
Line starting point

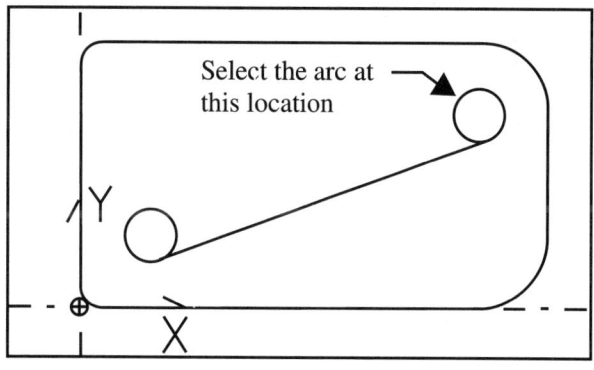

19. Check to verify that the **End** input field is highlighted. If it is not, select it with the mouse to highlight it.
20. Select the lower arc according to Figure 3.32.

At this time, your process model should match Figure 3.33.

The next step in the construction of your process model is to trim the unwanted areas of the arcs.

1. Choose **Geo Edit** from the workbench. If it is not on the workbench, it can be found under the **Edit** menu.
2. Choose **Trim/Extend**.
3. From within the **Trim/Extend** control panel, choose **Select 1st Element**.
4. Choose the lower arc as shown in Figure 3.34.
5. If the input field is not already highlighted, choose **Select 2nd Element**.
6. Choose the upper line as shown in Figure 3.35.

At this time, your process model should match Figure 3.36
To complete the trimming of the arc, follow these steps:

1. Again in the **Trim/Extend** control panel, highlight **Select 1st Element** by selecting it with your mouse.

**FIGURE 3.32**
Line Ending Point

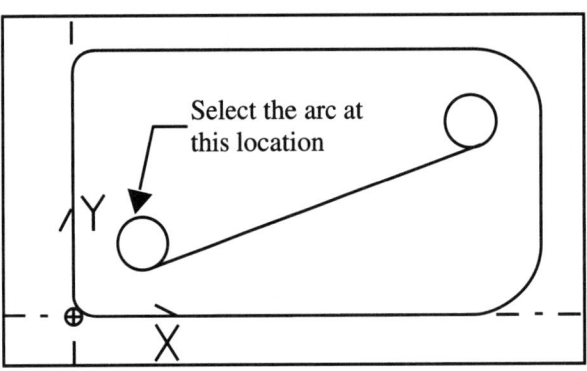

**FIGURE 3.33**
Completion of Tangent Lines

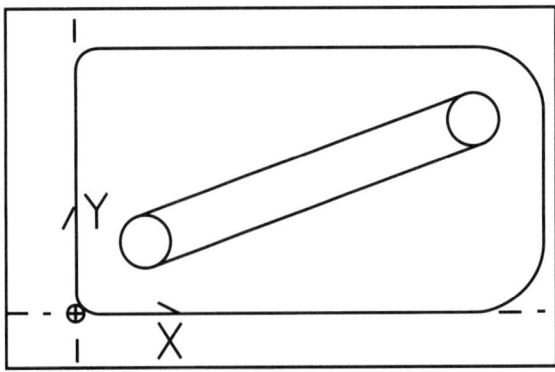

**FIGURE 3.34**
First selection point for trimming function

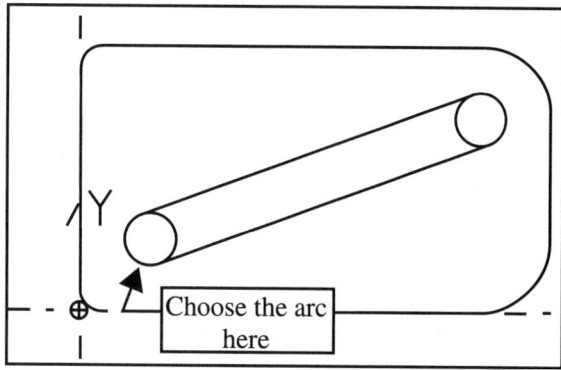

**FIGURE 3.35**
Second selection point for trimming function

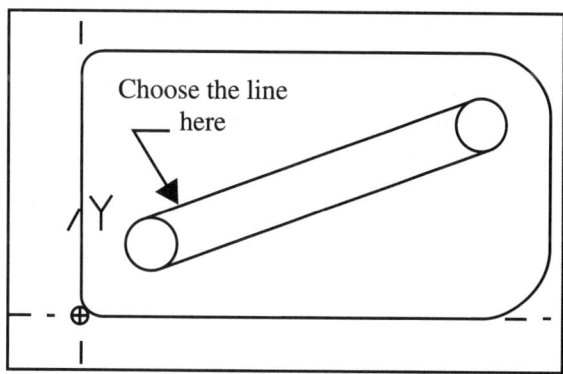

**FIGURE 3.36**
The partially trimmed arc

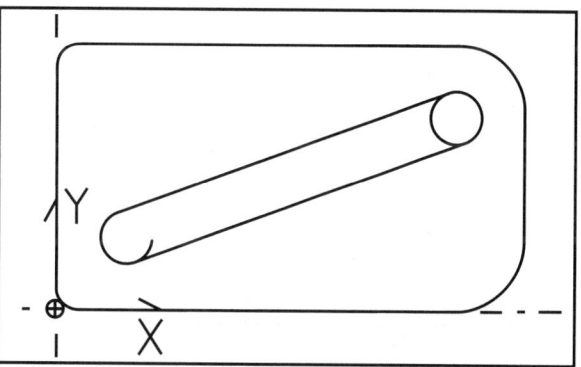

2. Choose the arc exactly as it was shown in Figure 3.34.
3. Highlight **Select 2nd Element** and choose the bottom line of the slot.

The entire arc should now be trimmed to the tangent point of the straight lines.

Continue with your process model by trimming the upper arc.

1. Again, highlight the **Select 1st Element** input field by selecting it with your mouse.
2. Choose the upper arc as shown in Figure 3.37.

**FIGURE 3.37**
First selection point to trim second arc

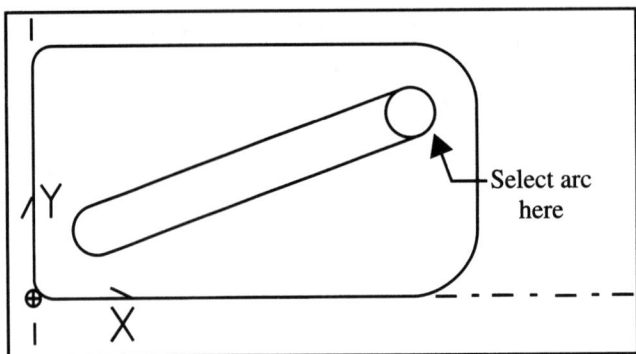

3. Verify that the **Select 2nd Element** input field is highlighted.
4. Select the bottom straight line of the slot.

The arc will trim as shown in Figure 3.38.

Continue with the process model by extending the arc to the upper straight line.

5. Verify that the **Select 1st Element** input field is highlighted.
6. Select the arc in the exact spot as was shown in Figure 3.37.
7. Select the upper straight line of the slot.

The arc will extend to the straight line and complete the construction of the finish profile as shown in Figure 3.39.

**FIGURE 3.38**
The upper arc partially trimmed

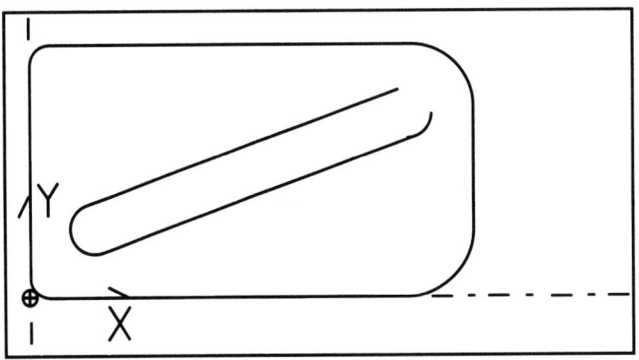

**FIGURE 3.39**
The completed finish profile

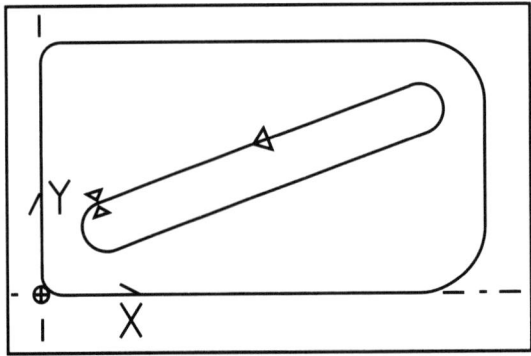

# Constructing the Roughing Profiles

Once you complete the construction of the finish profile, you can create the roughing geometry from the existing profiles. You must also resequence the database so that the roughing geometry is before the finish geometry.

The first task is to generate a roughing profile from the existing external finish profile. You must first identify the profile you wish to work with by making it the active group.

1. Select **Group** from the workbench.
2. Select **Profile** from the tool list.
3. Move your mouse to the external profile in the graphics work area and select any portion of the external geometry with your left mouse button. Grouping symbols which look like small arrowheads should appear on all eight elements of the external profile.

Next you must sequence your database so that the geometry will be machined in the correct order:

1. From the workbench, choose the **Insert** toolbox.
2. From the tool list, select the **Before** and the **Step Seq** modeling tools.

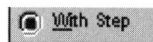

3. Choose the **With Step** option. Remember, a filled-in dot represents the "on" condition.
4. Fill in the input fields of the **Insert** control panel to match those of Figure 3.40.

   **Before Step** is set to 3, the finishing step.

   **With Step** is set to 1, the exterior profile roughing step.

   **Offset** is set to left so that the geometry will be created with a left offset on the cutting tool.

   **Level** is set to -.100", the depth of the first roughing pass.

   **Clear** is set to .100", the clearance point for the rapid traverse moves.

   **Profile Top** is set to 0.000", the top of the work piece.
5. From the workbench, choose **Geometry**.

**FIGURE 3.40**
The **Insert** control panel for step 1

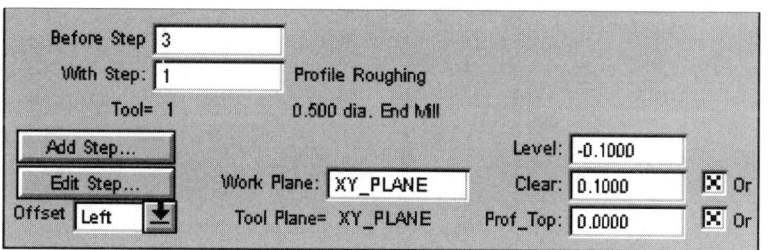

6. From the tool list, choose **Wall Offset**.
7. Fill in the input fields of your **Wall Offset** control panel to match those of Figure 3.41.

**Caution!** *You must fill in the input fields in the proper order to get the best results.*

8. Input the values for **Wall Side** and **Distance** first.
9. Highlight **Select Element in Profile**.
10. Choose the **Group Wall** button at the bottom of the control panel.

   **Wall Side** is set to Left. This will be to the outside of your work.

   **Distance** is set to an appropriate amount which you wish to leave on the workpiece for the finish tool to remove. In this example, .010" is left for finishing.

   **Wall Repeats** defaults to 1.

   **Corner Roll Angle** defaults to 180.00 degrees.

Upon selecting the **Group Wall** button, a roughing profile is created at a "Z" level of -.100". This roughing profile is exactly like the finish profile in every respect except that it is offset to the outside .010". At this time, your process model should match Figure 3.42.

The next process requires copying the roughing profile down four times at .100" per pass. To do this, follow these steps:

**FIGURE 3.41**
The **Wall Offset** control panel

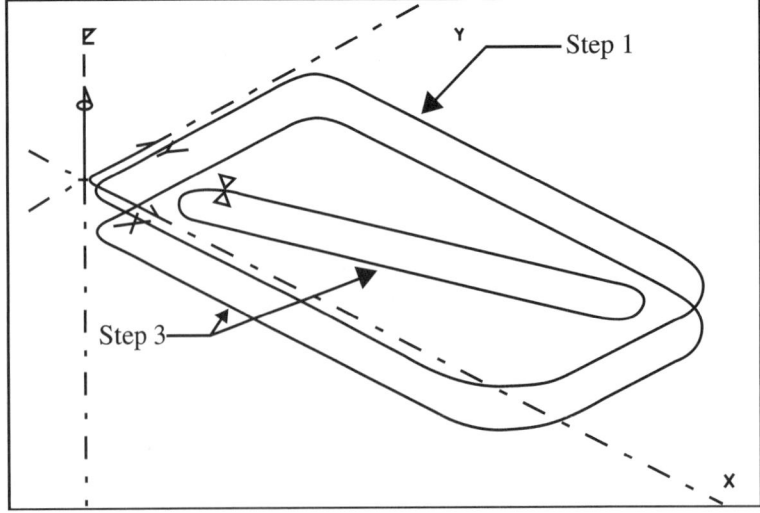

**FIGURE 3.42**
The first external Roughing Profile

1. From the workbench, choose **Group**.
2. From the tool list, select the **New Group** button. This will remove any active groups and allow the selection of a new group of elements.
3. From the tool list, select **Step**.
4. Inside the **Group** control panel, input 1 in the **Select Step** input field. This will make the roughing profile the active step. Arrowheads, which represent the active step, should now appear on the roughing profile.
5. From the **Edit** menu, choose **Transform**.
6. Choose **Move** from the tool list.
7. Fill in the input fields of your **Move** control panel to match those of Figure 3.43. Again, the order of selection is important.
8. First, turn **Copy** on by selecting the word.
9. Next, input 4 into the **Copies** input field. (Four copies plus the original will allow for five roughing passes at .100" each.)
10. For the **From Point** input field, simply choose the **From 0** button.
11. For the **To Point** input field, input "X" 0.000, "Y" 0.000, and "Z"-.100.

This will complete the external roughing passes. At this time your process model should match Figure 3.44.

**FIGURE 3.43**
The **Move** control panel

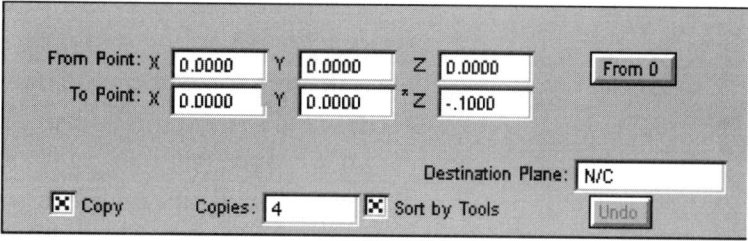

**FIGURE 3.44**
The completed external roughing passes

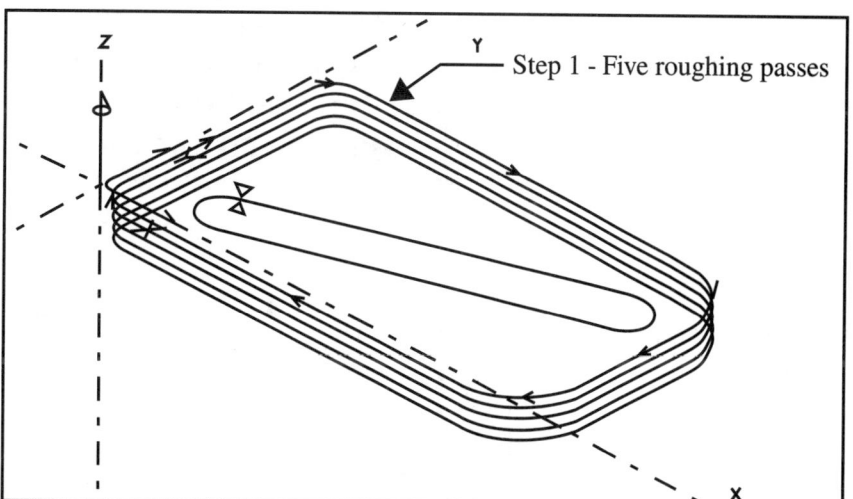

# Constructing the Roughing Passes for the Slot

The next step in the construction of your process model is creating the geometry necessary to rough the slot. Recall that during the construction of the job planner, you assigned the same tool to be used for step 1 and for step 2. Step 2 will be used to rough the slot. This will allow you independent control of the speeds and feeds while using the same tool as was used to rough the exterior profile.

1. Choose **Group** from the workbench.
2. Choose the **New Group** button. Remember, the selection of this button will deactivate an active group.
3. Choose the **Profile** tool from the tool list.
4. With your mouse, select any element which is a part of the finish profile of the slot. You should now see grouping symbols (narrow arrowheads) on the profile of the slot.
5. From the workbench, select the **Insert** toolbox.
6. Select the **After** tool and the **Step Seq** tool from the tool list.
7. Fill in the input fields of your Insert control panel to match those of Figure 3.45.

   **After Step** is set to 1. The slot roughing geometry needs to come right after the exterior profile roughing to avoid excessive tool changes.

   **With Step** is set to 2—the slot roughing step.

   **Offset** is again set to Left so that the next piece of geometry that is created will have a left offset on the cutting tool.

   **Level** is set to -.050"—the depth of cut for the first roughing pass.

   **Clear** is set to .100".

   **Profile Top** is set to 0.000".
8. Choose **Geometry** from the workbench.
9. Choose the **Wall Offset** modeling tool.
10. Fill in the input fields of your **Wall Offset** control panel to match those of Figure 3.46. Remember, the order of input is important in this control panel.

**FIGURE 3.45**
The **Insert** control panel for the slot

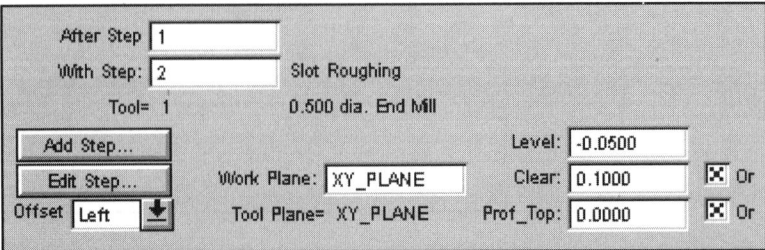

**FIGURE 3.46**
The **Wall Offset** control panel

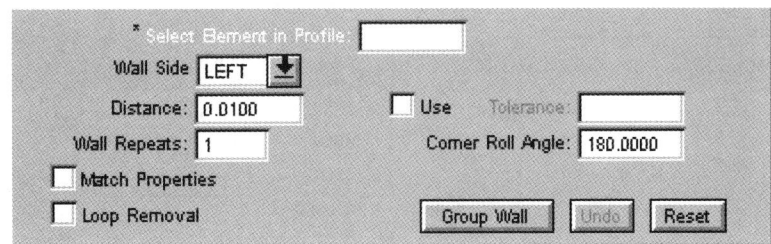

**Wall Side** is set to Left—the inside of the pocket.

**Distance** is again set to .010″ which is an appropriate amount for a finish cut.

After these two input fields are complete, highlight **Select Element in Profile** and choose the **Group Wall** button.

The first pass of the slot roughing geometry should now be shown and should match Figure 3.47.

11. From the workbench, choose **Group**.
12. From the tool list, choose the **New Group** button.
13. Choose the **Step** tool.
14. Select step 2 from the database list.
15. Choose the **Transform** toolbox from the workbench. (If **Transform** is not on the workbench it will be under the **Edit** menu.)
16. Choose the **Move** tool from the tool list.
17. Fill in the **Move** control panel to match Figure 3.48. (Remember to fill in the Copy input field first, followed by **Copies,** then **From Point,** and last **To Point**).

**Copy** is turned on as signified by the "X" in the box.

**Copies** is set to 4—the original at -.050″ plus 4 more passes at -.050″ will equal the .250″ depth of the slot.

**FIGURE 3.47**
The first roughing pass of the slot

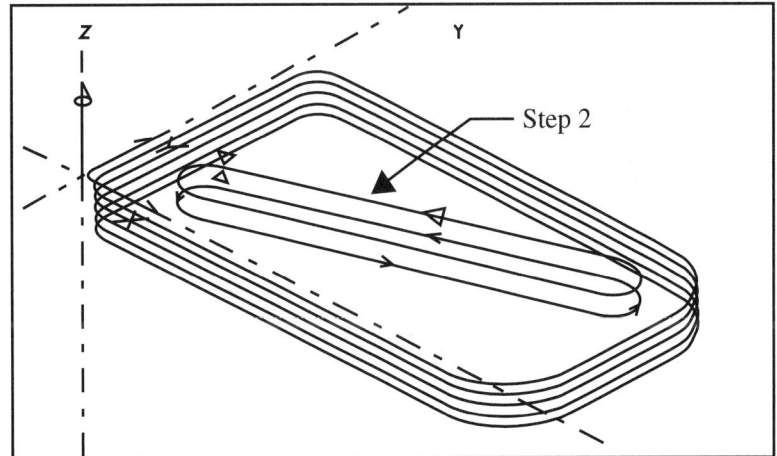

**FIGURE 3.48**
The **Transform** control panel

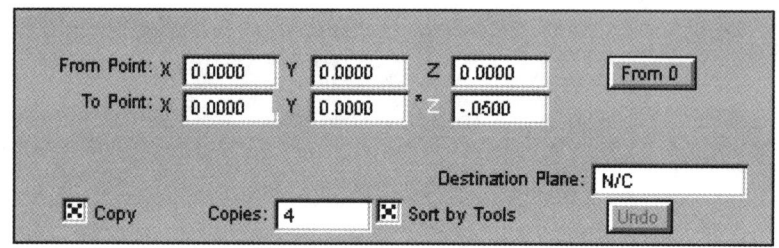

> **From Point** is set to 0.000 for all input fields. Simply highlight the words "From Point" and select the **From 0** button.
>
> Highlight the "X" input field of the **To Point** selection and input 0.000.
>
> Highlight the "Y" input field of the **To Point** selection and input 0.000.
>
> Highlight the "Z" input field (notice the trigger). Input -.050 and press the enter key of the keyboard.

At this point the process model is complete and should match Figure 3.49.

**FIGURE 3.49**
The completion of the process model

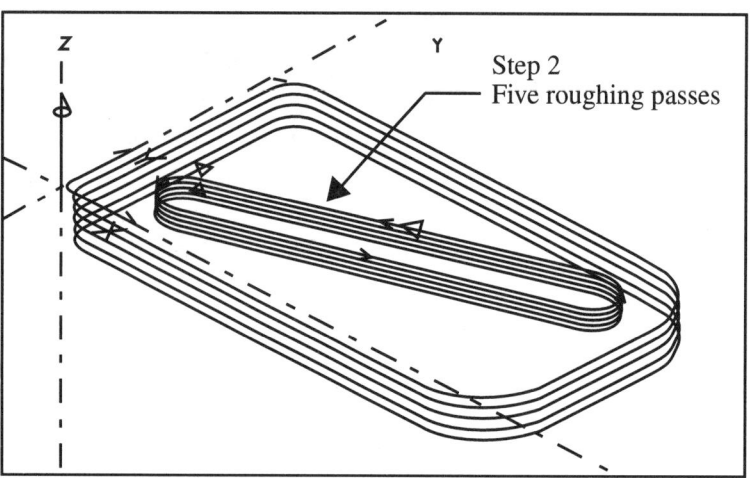

# CHAPTER 4

# SmartCAM Tutorial 3

## Pocket Mill/Drill

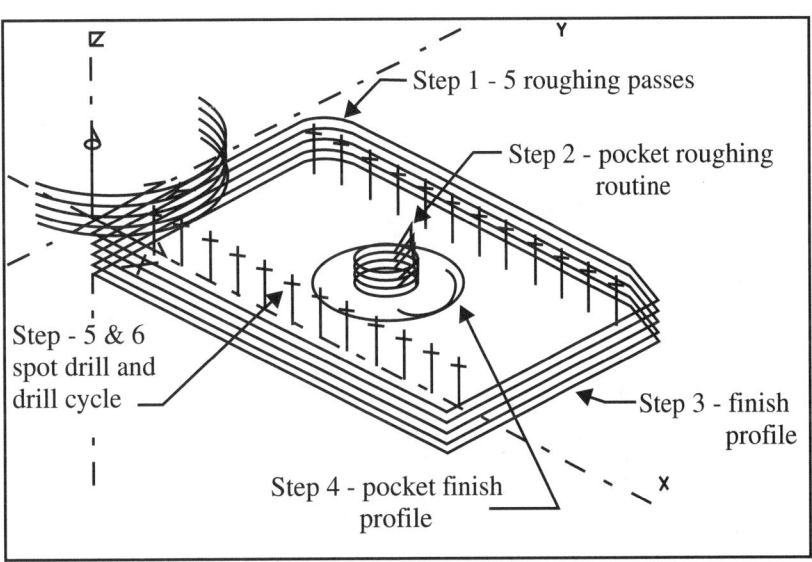

## Chapter 4 SmartCAM Tutorial 3

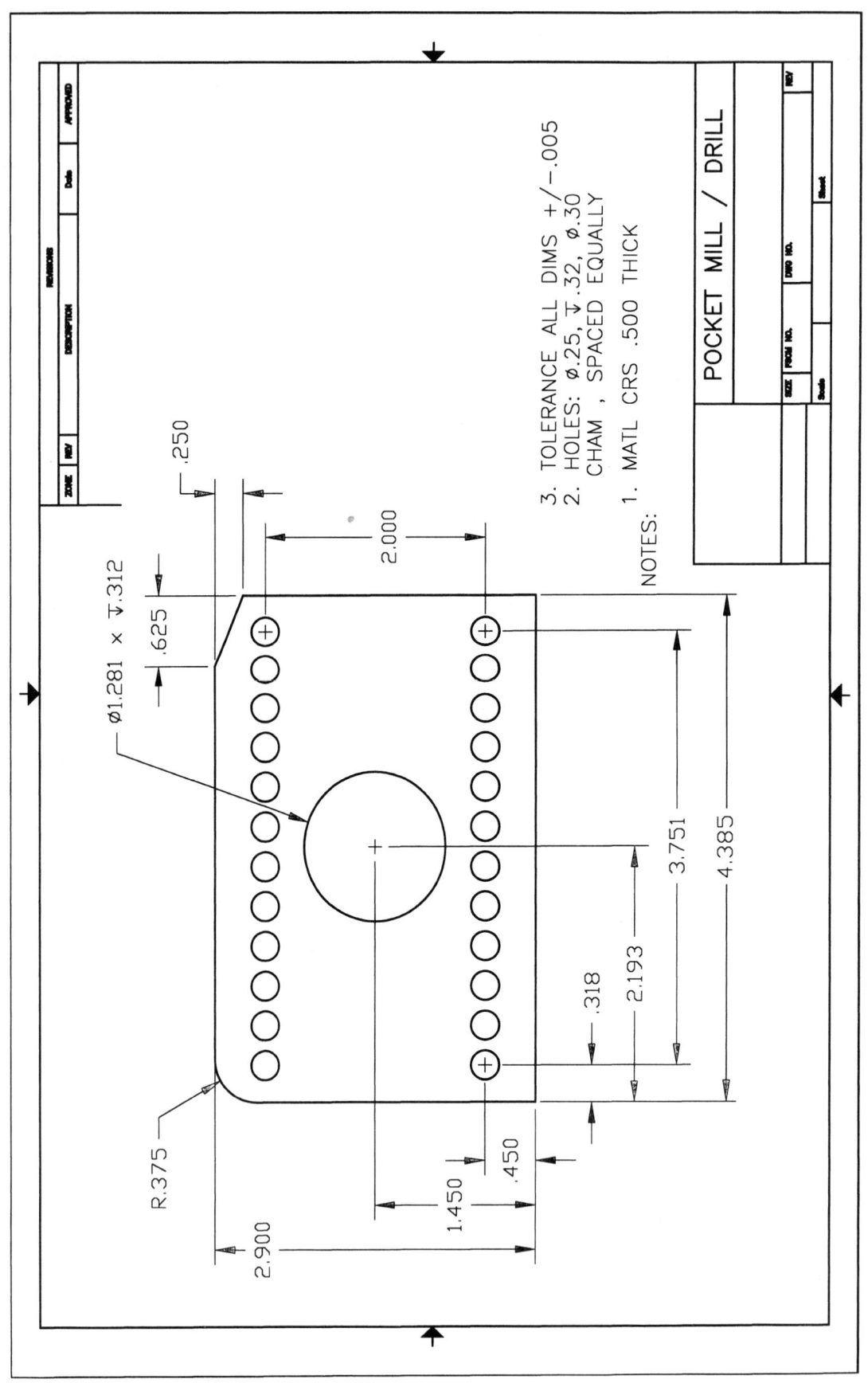

The blueprint for the Pocket Mill/Drill Tutorial

Chapter 4 SmartCAM Tutorial 3

Upon completion of this chapter, you should be able to:

- Construct geometry which consists of lines and arcs.
- Successfully use SmartCAM's Blend and Chamfer functions.
- Add Lead In and Lead Out geometry to enhance the machining process.
- Split an element to facilitate the placement of Lead In/Lead Out geometry.
- Understand and apply SmartCAM's pocket roughing function.
- Use SmartCAM's hole function to successfully create a drilled hole.
- Transform a single hole into a multiple hole pattern.

# POCKET MILL/DRILL

The third tutorial in this text will introduce you to the concepts of adding a chamfer to your external profile, building a pocket roughing routine, and spot drilling and then drilling a hole pattern. Additionally, geometry that allows you to lead into and out of your work will be added to all relevant profiles to more closely parallel industrial machining applications.

The first operation required in the construction of project 3 is to build a Job File. Create a job file that includes the steps and tools shown in Table 4.1. If necessary, refer to the previous tutorial to review the proper steps to construct a Job Operations File.

## Constructing the Finish Profiles

After completion of the Job Operations File, you may begin building your process model. The first step of your process model is building the finish toolpath of the exterior profile.

**TABLE 4.1**

| Step # | Tool # | Type | Diameter | Speeds | Feeds |
|---|---|---|---|---|---|
| 1 | 1 | End Mill (2 flute roughing) | 3/4" | 80 SFPM | .004 IPT |
| 2 | 1 | End Mill (2 flute roughing) | 3/4" | 80 SFPM | .004 IPT |
| 3 | 2 | End Mill (4 flute finishing) | 3/4" | 90 SFPM | .003 IPT |
| 4 | 2 | End Mill (4 flute finishing) | 3/4" | 90 SFPM | .003 IPT |
| 5 | 3 | Spot Drill | 5/16"(90 deg) | 90 SFPM | 11 IPM |
| 6 | 4 | Drill | 1/4" | 90 SFPM | 10 IPM |

To build the finish toolpath of the exterior profile, follow these steps:

1. Choose **Insert** from the workbench.
2. Choose **After, Element sequence** and **With Step** from the tool list.
3. Fill in the input fields of your **Insert** control panel to match those in Figure 4.1.

   **After Element** defaults to element number 0. At this point of your process model there are no other elements in your database list.

   **With Step** is set to 3, the finish milling step. The blueprint reflects finish geometry. You therefore must construct your finish profile first

   **Offset** is set to Left. The external profile of your model will be constructed in a clockwise manner. A left offset will allow you to climb mill the workpiece.

   **Work Plane** is set to the XY_Plane for this model.

   **Level** is set to -.500, the depth of your finish geometry. Remember, you are constructing your finish profile first. The finish tool will cut one final pass around to take out any steps left by the roughing tool.

   **Clear** is set to .100. This is the value to which the tool will retract prior to any rapid traverse positioning moves.

   **Profile Top** is set to 0, the top surface of your model.

In order to create the geometry of the process model, the **Geometry** toolbox must be on the workbench. If it is not, it is located under the **Create** menu. To begin the construction of the finish geometry, follow these steps:

1. Choose **Geometry** from the workbench.
2. Choose the **Line** tool from the tool list.
3. Fill in the input fields of your **Line** control panel to match those in Figure 4.2.
4. Continue creating lines until you have a completed rectangle which measures 4.385" wide by 2.9" high.

**Hint:** *When you are filling in the input fields of the **Line** control panel, you can input values with your keyboard or, once you have an element in your working area, you can use your mouse and "snap to" existing elements.*

**FIGURE 4.1**
The input fields of the **Insert** control panel

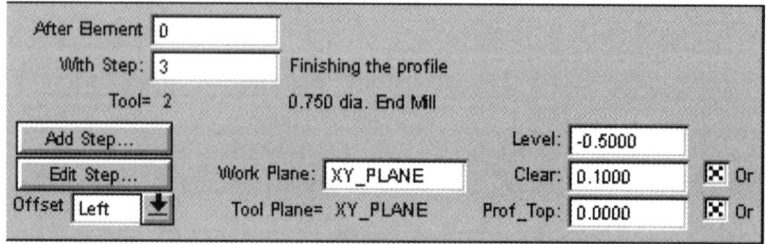

Chapter 4   SmartCAM Tutorial 3

**FIGURE 4.2**
The input fields of the **Line** control panel

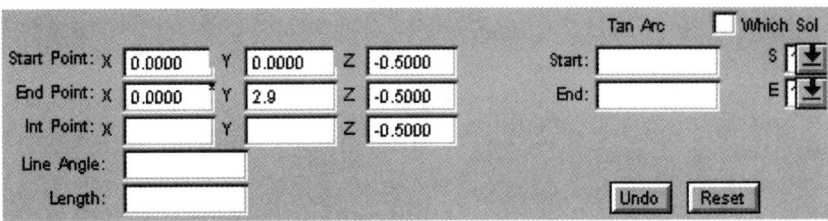

To demonstrate this procedure:

1. Draw the first line as per the control panel in Figure 4.2.
2. Make sure the **snap** function is turned on. The snap icon appears as a button that has been depressed when it is on.
3. Make sure the endpoint function is also on.
4. With your mouse, highlight the word **Start Point** of the **Line** control panel.
5. Pick the end point of the element that was created by the completion of the control panel of Figure 4.2.

Notice how all three values of the **Start Point** input field are simultaneously entered by the snap function? The same results can be achieved if you need to extract only the "X," "Y," or "Z" values. Simply highlight the appropriate field (x, y, or z), and use your mouse to snap to the endpoint of the appropriate element.

6. Continue by entering the appropriate **End Point** from your keyboard.

**Remember:** *SmartCAM will machine the elements in the numerical order reflected in your database list. They will also be machined in the direction in which they were created. You will want to create this rectangle in a clockwise manner, starting from the origin. Your completed rectangle should match Figure 4.3.*

To place the 3/8" radius on the upper left corner of the model you will use the **Geo Edit** toolbox. If it is not already on the workbench, it can be found under the **Edit** menu.

1. Choose **Geo Edit** from the workbench.
2. Choose the **Blend** tool from the tool list.

**FIGURE 4.3**
The completed rectangle

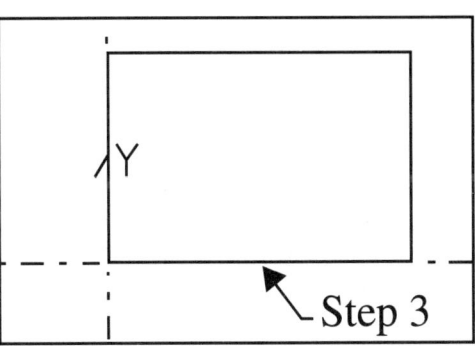

3. Inside the Blend control panel, set the **Inside Radius** input field to .375.
4. With your mouse, highlight **Select 1st element**.
5. Choose the first element of the database (this should be the left vertical line). SmartCAM will allow you to pick the element using your mouse, from the graphics work area, from the database list, or you can type the number of the element directly into the input field.
6. Next, choose **Select 2nd element** and choose the number 2 element of the data list (this should be the top horizontal line).

At this time you should see the 3/8" radius in the upper left corner of the part, as shown in Figure 4.4.

The next step in the construction of your external profile is placing the chamfer on the upper right corner of the model.

To place the chamfer on the upper right corner of the model, follow these steps:

1. Choose **Geo Edit** from the workbench. (It is found under the edit menu if it is not on the workbench.)
2. Choose **Chamfer** from the tool list.

Upon selecting **Chamfer** from the tool list, the **Chamfer** control panel will open at the bottom of your screen. You must complete the input fields in the proper order to get the desired results.

1. With your mouse, highlight the **Parallel Size** input field.
2. Input .625.
3. With your mouse, highlight the **Perp. Size** input field.
4. Input .250.
5. With your mouse, highlight **Select 1st Element**.

**FIGURE 4.4**
The 3/8" blended radius

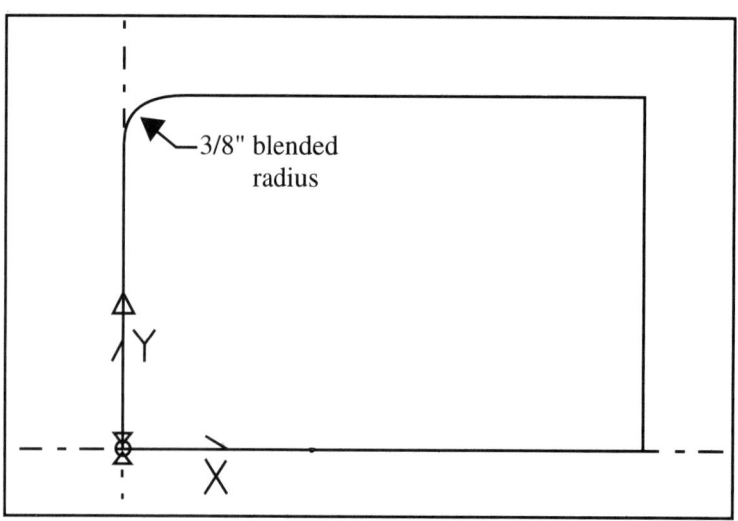

6. Select the top horizontal line from the graphics work area.
7. With your mouse, highlight **Select 2nd Element**.
8. Select the right vertical line from your graphics work area.

Figure 4.5 shows the Chamfer control panel along with the correct values in each input field.

**Parallel Size** refers to the dimensional value of the chamfer that is parallel to the first element chosen.

**Perp. Size** refers to the dimensional value of the chamfer that is perpendicular to the first element chosen.

**Caution:** *During the course of working with this control panel, a minor problem occasionally arises. If the operator fails to properly enter the input field values, SmartCAM will display the error message shown in Figure 4.6.*

On other occasions the **Select 2nd Element** input field has simply been grayed out and unavailable for selection. When you select the **Parallel Size** input field, input the value .625 and make sure you hit the enter key. Likewise, when you select the **Perp. Size** input field, input the value .250 and hit the enter key. Failure to properly enter the values by simply picking the input fields with your mouse will result in problems.

At this point, your process model should match Figure 4.7.

**FIGURE 4.5**
The Chamfer control panel

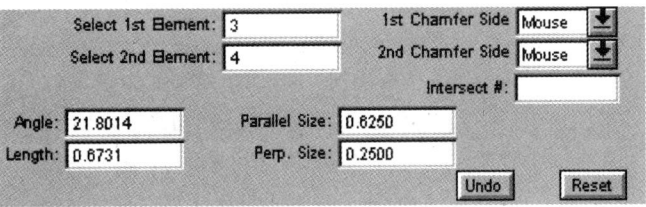

**FIGURE 4.6**
Potential error message

**FIGURE 4.7**
The completed external profile

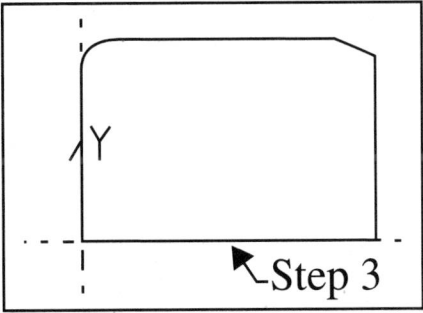

# Adding Lead In/Lead Out Moves

If you watch the plunge movement of the tool during **Showpath** (to do so choose View and then choose Showpath) you will observe the endmill plunging at a point tangent to the finish profile. During any plunge move of an endmill, tool deflection will cause a gouge if the plunge location is too close to the finish profile. This tool deflection will result in defective parts. The solution is to add a lead in/lead out move to your finish profile. A lead in move will allow the tool to move to a specified distance away from the profile, plunge to depth and then move into the work.

To properly add a lead in/lead out move to your profile, follow these steps:

1. Choose **Geo Edit** from the workbench.
2. Choose **Lead In/Out** from the tool list.
3. Fill in your **Lead In/Out** control panel to match Figure 4.8.

This example will use **Both** a lead in move and a lead out move.

4. Choose only an **Arc** type lead in/lead out move.

   The **Angle** input field, for this example, should be something other than 90 degrees.

   An angle that positions the start and the endpoints of a profile at the same coordinates, such as a 90 degree angle would do in this example, will cause problems on **Showpath** and on **Code**.

   The **Radius** field value should be greater than that of the tool.

   The **Select Element in Profile** should be the first element of the profile, in this example element 1.

With the addition of the lead in/lead out moves, your process model should now match Figure 4.9.

In some applications, the position of the lead in/lead out move in the previous figure may not be desirable due to the position of vise jaws or some other obstruction. Because that may be a factor, reposition the lead in/lead out move to the center of the left vertical line.

1. If the **Lead In/Out** control panel is still open, choose undo. (If the undo function is grayed out and unavailable, choose **Geo_Edit > Delete** and delete the lead in/lead out move.)
2. Choose **Geo Edit** from the workbench.

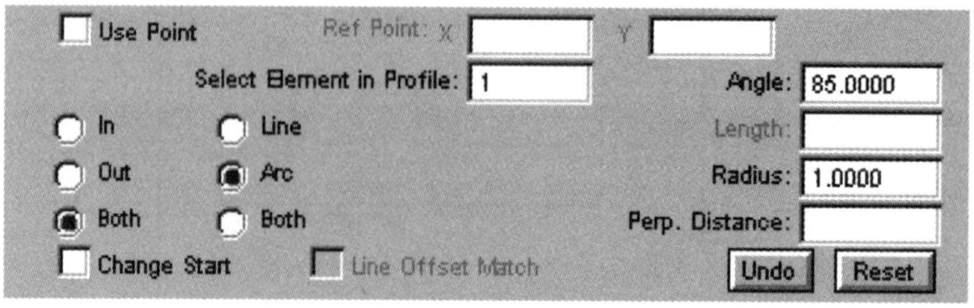

**FIGURE 4.8**
The **Lead In/Out** control panel

Chapter 4   SmartCAM Tutorial 3

**FIGURE 4.9**
The inserted lead in/lead out geometry

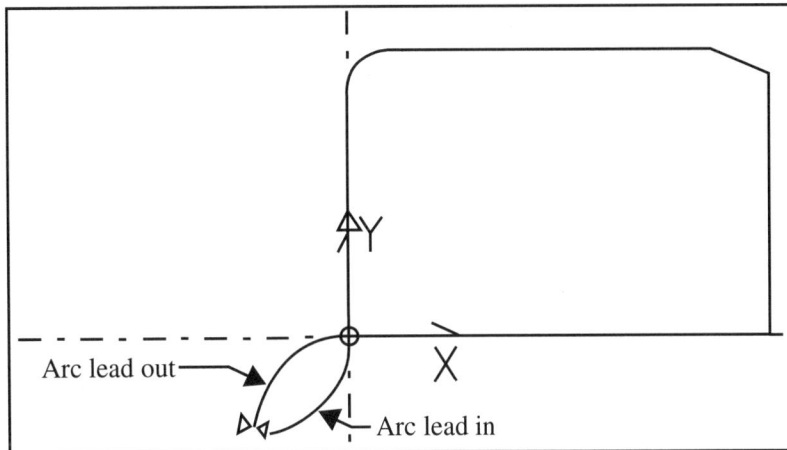

3. Choose **Split** from the tool list. Split will allow you to divide the left vertical line into two elements.
4. From within the Split control panel, choose **Element Division.**
5. For **% Length**, accept the default value of .5.
6. Choose **Select Split Element.**
7. Choose element 1 (the left vertical line).

If you look closely at the line you will notice a small asterisk at the point the line was split. If you do not see an asterisk, DO NOT attempt to split the line again. You may wind up with multiple splits. Complete the exercise before determining if the line was split correctly.

8. Choose **Lead In/Out**.
9. Turn the **Change Start Point** selector switch on. This is necessary to allow you to resequence the database. Lead in/lead out moves can only be placed on the first element of a profile. Currently the vertical line on the left, which starts at the origin, is the first element of the database. The vertical line above the asterisk is the one that needs to be the first line of the database.
10. Set the remaining input fields to the values shown in Figure 4.8, with the exception of **Select Element in Profile**.
11. For the Select Element in Profile input field, choose the left vertical line above the asterisk (above the midpoint) but below the 3/8" arc.

With the completion of the lead in/lead out geometry, the external profile of your process model is complete and should match Figure 4.10.

## Construction of the Pocket Profile

The next operation to be performed is the construction of the 1.281" diameter hole in the center of the part. Since the profile depth is changing and the step

**FIGURE 4.10**
The completed external finish profile with the modified lead in/lead out arcs

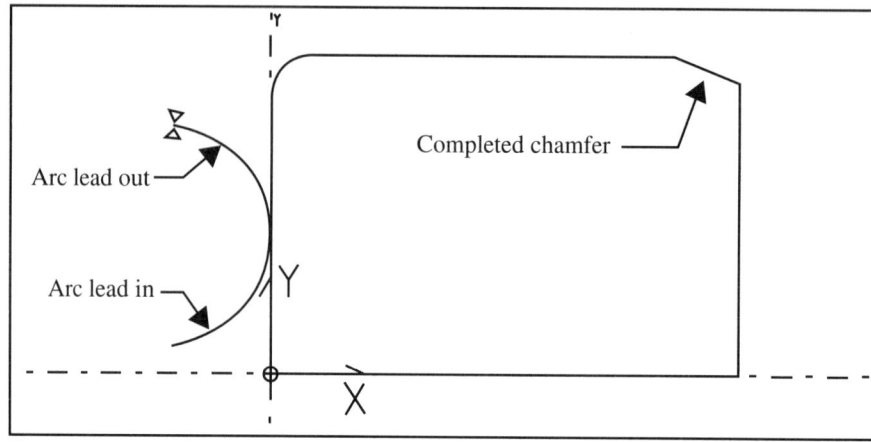

properties are changing, you must return to the **Insert** control panel to activate these changes.

1. Choose **Insert** from the workbench.
2. Choose the **After, Step Seq**, and the **With Step** options from the tool list.
3. Fill in the input fields of your **Insert** control panel to match those in Figure 4.11.

Once the input fields of the **Insert** control panel are correctly filled in, you will need to construct the appropriate geometry.

1. Select **Create** from the menu bar.
2. Choose **Geometry** from the menu.
3. Select **Arc** from the tool list.
4. Fill in the input fields of your **Arc** control panel to match those in Figure 4.12.

**FIGURE 4.11**
The **Insert** control panel for step 4

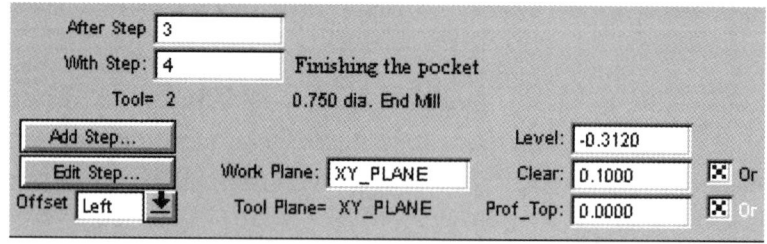

**FIGURE 4.12**
The **Arc** control panel

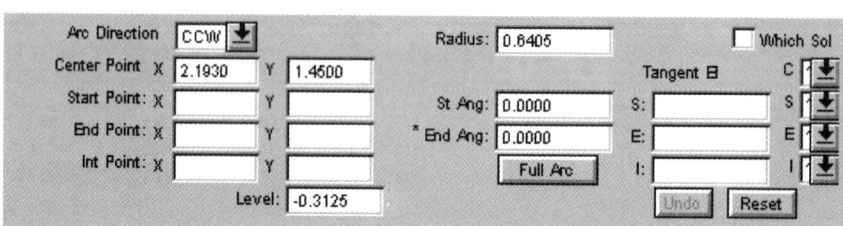

**Arc Direction** should be set to CCW (counterclockwise). Remember, the **Offset** input field of the **Insert** control panel was set to left. This was done so that you can climb mill your profile. You are now on the **inside** of your profile, therefore your geometry must be constructed in a counterclockwise manner.

**Radius** is set to the radius of the hole. Notice that the blueprint dimensions the hole as per the diameter. SmartCAM will accept a mathematical formula as input. Enter 1.281/2 into the **Radius** input field. SmartCAM will calculate the result when you enter the formula.

**Center Point** is the center of the hole in the "X" and the "Y" coordinates. For the "X" input field, enter 4.385/2. For the "Y" input field enter 2.9/2.

At this point in the process model SmartCAM has enough information to show you a potential solution to your arc. Notice the dashed representation of the hole. All you need to do now is specify the **Start Point** and the **End Point** of the arc. Since the arc is a complete circle, select the **Full Arc** button with the mouse. SmartCAM will replace the dashed representation with a solid line.

Notice also the **Level** input field. The -.312" value was carried over from the **Level** input field of the **Insert** control panel.

At this point, your process model should match the isometric view of Figure 4.13.

**Caution:** *Do not put a lead in/lead out move on the pocket profile at this time. This must be done **after** the pocket roughing routine is complete. Doing so will cause SmartCAM to display an error message stating that the outside boundary must be closed.*

# Constructing the Roughing Profiles

After completion of the finish geometry, the next step is to use the finish profiles to generate the roughing profiles. Again, open the **Insert** control panel since most of your step properties will change for the next operation.

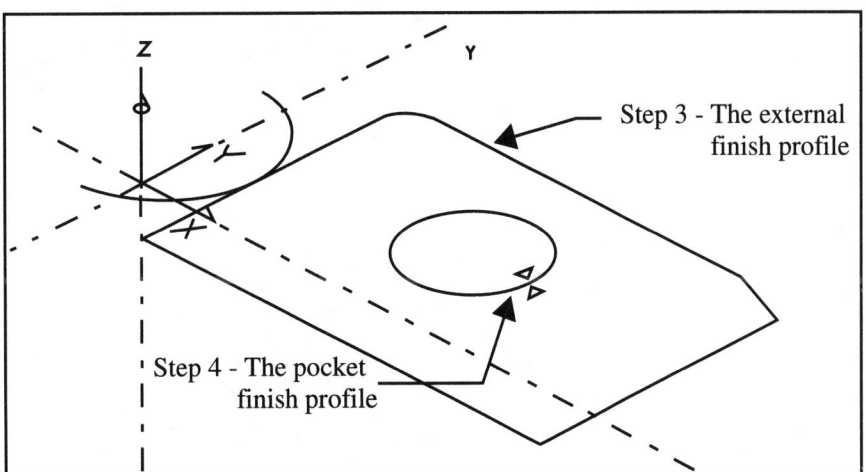

**FIGURE 4.13**
The external finish profile with the pocket finish profile

1. Select **Insert** from the workbench.
2. Select the **Before, Step Seq**, and the **With Step** options from the tool list.
3. Fill in the input fields of your **Insert** control panel to match those in Figure 4.14.

Again, the purpose of this control panel is to resequence the database list to place all geometry created with step 1, the roughing tool, to come before step 3, the finishing tool. The geometry is also placed at a level of -.100", the depth of your first roughing cut.

You are now ready to group the exterior profile.

1. Choose **Group** from the workbench.
2. Select the **Step** option.
3. Choose step 3, the exterior profile.

Once the Group-by-Step control panel is open, you have the option of picking the geometry from the graphics work area or the database list. (The **With Step** selector switch must be on to view the list of steps in the database list.) Additionally, simply typing a "3" in the **Select Step** input field of the Group control panel will achieve the same results.

If you recall from previous lessons, the purpose of grouping is to identify certain elements with which you wish to work. When elements are part of an active group, an arrowhead will be displayed on the center of the element, and a dash will be displayed beside the element number in the element data list.

Once the elements are grouped, you can then create the roughing profile. To create the roughing profile, follow these steps:

1. Choose **Geometry** from the workbench.
2. Choose **Wall Offset** from the tool list.
3. Fill in the input fields of your **Wall Offset** control panel to match those in Figure 4.15.

**FIGURE 4.14**
The **Insert** control panel

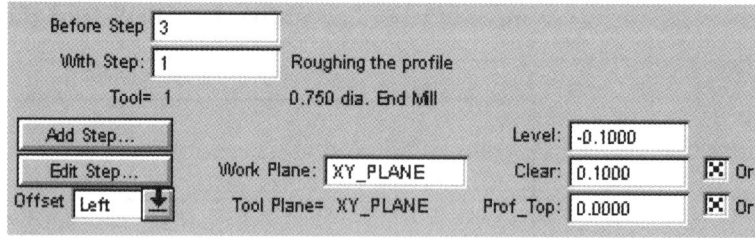

**FIGURE 4.15**
The input fields of the **Wall Offset** control panel

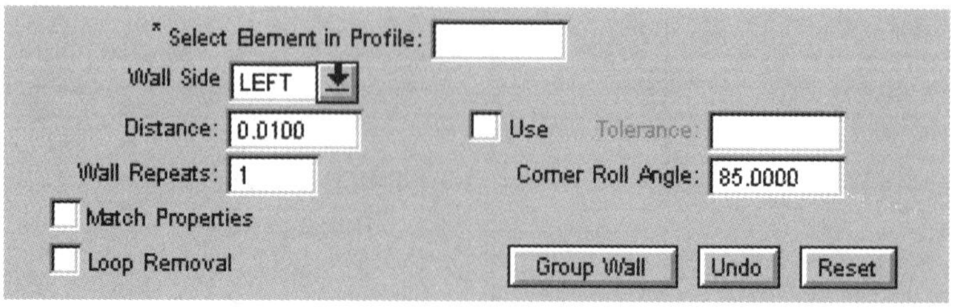

**Wall Side** is set to left. You will want your roughing profile to be on the outside of the finish cut. To determine which side of the element is right or left, imagine you are standing on the line to be cut, facing the direction in which the cutter will feed. Left offset will be according to your left, and right offset will be according to your right.

**Distance** is the perpendicular distance you want to maintain between the finish profile and the roughing profile. In other words, after the roughing passes are complete, how much material do you want to leave on for the finish tool?

**Wall Repeats** is set to one. You only want one offset for this part. If the type of material you were cutting required more than one roughing pass for each "Z" level cut, you have the option of increasing this value.

**Corner Roll Angle** is set to a value less than the angle between intersecting elements. In this example, all straight line intersecting points of the process model have an angle of less than 180 degrees. With the **Corner Roll Angle** input field set to the 180 degree default setting, SmartCAM will automatically insert a corner blend at each of these intersection points. To prevent this blend from occurring, the value must be less than any angle in the process model.

All other fields are left blank signifying they are off.

4. Choose the **Group Wall** button.

At this point, your process model should match Figure 4.16.

The next step in creating your roughing geometry is to copy four additional roughing passes in the "Z" axis.

Group the roughing profile:

1. Choose **Group** from the workbench.
2. Select the **New Group** button to deactivate the active group.

**FIGURE 4.16**
The result of the Wall Offset function

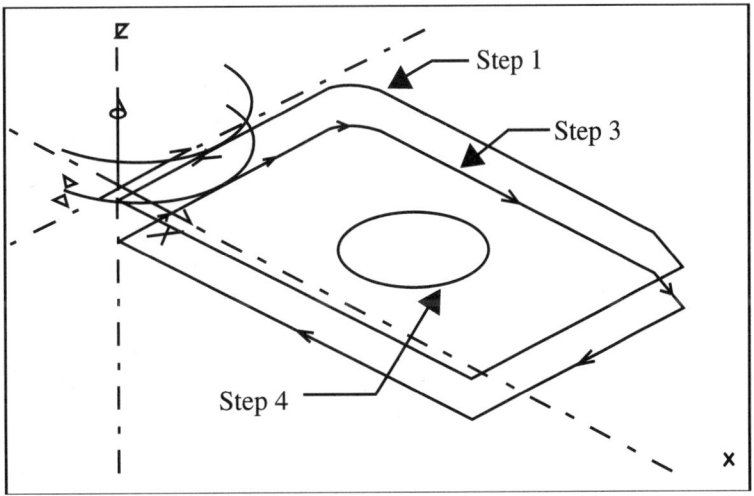

3. Select **Step** and select Step 1.
4. Choose **Edit** from the menu bar.
5. Choose **Transform**.
6. Choose **Move**.

Fill in the input fields of your **Move** control panel to match those in Figure 4.17.

**Copy** should be turned on as the first step in this control panel.

**Copies** is set to 4 (the original, plus 4 copies, all at an incremental "Z" depth of -.100" will equal the total depth of -.500").

For the purposes of copying grouped geometry in the "Z" axis, you need the **From Point** input field to contain all zero values. Simply pick the "From 0" button.

**To Point** should have the "X" and the "Y" values set to zero and the "Z" value the INCREMENTAL distance from one piece of geometry to the next.

**Remember:** *In this example, this control panel should be viewed as incremental coordinates. There are other situations where the values in this control panel will reflect the absolute coordinates. However, this is not one of them.*

At this point, your process model should match Figure 4.18.

**FIGURE 4.17**
The input fields of the **Move** control panel

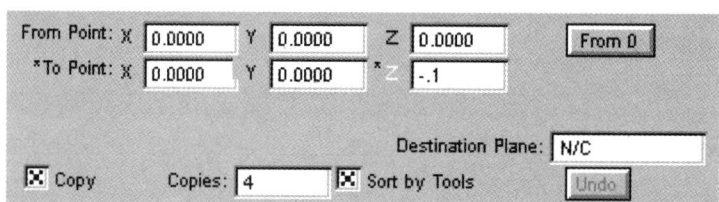

**FIGURE 4.18**
The completed external roughing geometry

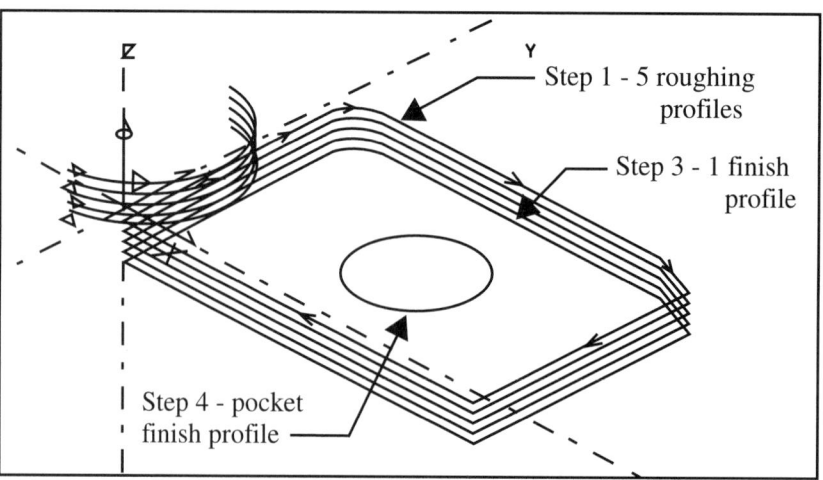

Chapter 4 SmartCAM Tutorial 3                                                                 75

# Roughing the Pocket

The next step of this process model will introduce you to SmartCAM's pocket roughing routine. This pocket roughing option will automatically remove material from a **closed** pocket. "Closed" means that the endpoints of the geometry must touch. In addition, it is a good idea to verify that the profile top of the circle is set to zero and the geometry is at the proper depth of -.312".

Verify these values before proceeding:

1. Choose **Utility** from the menu bar.
2. Choose **Element Data**.
3. With your mouse, pick the circle which defines the pocket.

A read only dialog box will open revealing all properties for the selected element.

To correctly construct the pocket roughing profiles, you first need to resequence the database so that the roughing cuts will come before the finish cuts.

1. Choose **Insert** from the workbench.
2. Select the **After, Step Seq**, and **With Step** options from the tool list. Fill in the input fields of your **Insert** control panel to match those in Figure 4.19.
3. Obtain a full-scale, top view of the model.
4. Choose **Process** from the menu bar.
5. Choose **Rough** from the menu.
6. Choose **Pocket** from the tool list.

Fill in the input fields of your **Pocket** control panel to match those in Figure 4.20.

**FIGURE 4.19**
The **Insert** control panel

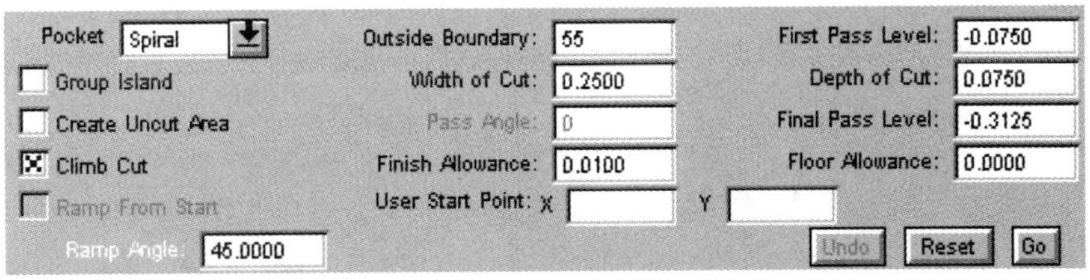

**FIGURE 4.20**
The input fields of the **Pocket** roughing control panel

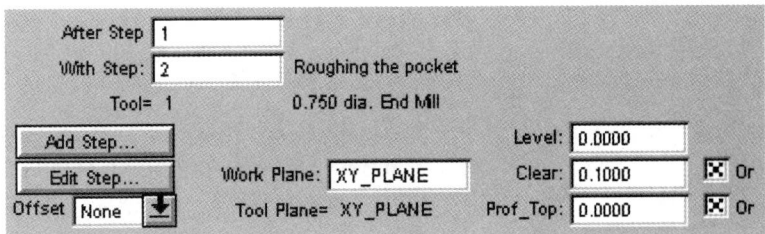

**Pocket** should be set to spiral. Other options such as zigzag and linear are available, however, spiral works best for this example.

**Climb Cut** is turned on (an "X" in the box indicates the switch is on). The roughing routine will work equally as well with this switch turned off, however, the geometry will be generated in a counterclockwise manner with **Climb Cut** turned on.

**Ramp Angle** is set to 45 degrees. This is the angle at which the tool will ramp into the work. Other angles can be used, and the end result will be the same. Smaller values will cause a smaller approach angle. This is easier on the cutter, however, it may slow the production rate. Smaller values also require more room for the ramp. Larger values up to 90 degrees will result in a more direct plunge of the cutter. This will be harder on the cutter (and machine) and may not be desirable for tougher materials.

**Caution:** *Make sure you use a center cutting endmill for an operation such as this.*

**Outside Boundary** is the element that defines the pocket. In this example it is the circle. Simply pick the circle with your mouse.

**Width of Cut** defaults to .375". This is half of the cutter diameter that was entered during the building of the Job Plan. Other values will work with the same end result. Smaller values will cause smaller stepover amounts and will be easier on your cutter. Larger values will increase your production rate, but at the same time will be harder on the cutter and machine.

**Pass Angle** in not an option with the Spiral pocket routine. Remember, in SmartCAM when a selection is "grayed out," that choice is not available for you to pick. When either Linear or Zigzag is chosen as the pocket routine, **Pass Angle** will be in bold print. The value will be the angle of the start point of the cut.

**Finish Allowance** is the amount of material you want to leave on the wall of the pocket for the finish tool to remove.

**User Start Point** is left blank for this example. This input field will indicate where the cutting routine is to begin. In a pocket routine such as this, it is better to let the system decide where to start the cut. Other operations, such as face milling will better utilize this option.

**First Pass Level** is the "Z" level at which the first roughing cut is to take place. This must be a negative value.

**Depth of Cut** is the "Z" level at which all successive cuts are to take place. This must be a positive value.

**Final Pass Level** is the "Z" level at which the full depth of the pocket resides. This must be a negative value.

Chapter 4   SmartCAM Tutorial 3

**Floor Allowance** is set to zero for this example. This is the amount of material which you would leave on the floor of the pocket for your finish tool.

When these input fields are set correctly in your control panel, select the "Go" button with your mouse.

At this time, your process model should match Figure 4.21.

Upon completion of the pocket roughing routine, the proper placement of the lead in/lead out move on the pocket finish profile is possible. Attempting to place the lead in/lead out move on the pocket profile prior to the pocket roughing routine, will result in the error message shown in Figure 4.22.

The reason for this error message is really very logical. SmartCAM analyzes the profile and recognizes that, with a lead in/lead out arc, the starting point of the lead in arc is not the same coordinate value as the endpoint of the lead out arc. The profile, therefore, cannot be closed.

The simplest way around this error message is to construct the lead in/lead out arcs of a finish pocket after the pocket roughing process is completed.

To create the lead in/lead out arcs of the pocket profile:

1. Choose **Edit** from the menu bar.
2. Choose **Geo Edit** from the menu.
3. Choose **Lead In/Out** from the tool list.

Fill in the input fields of your **Lead In/Out** control panel to match Figure 4.23.

**FIGURE 4.21**
The completed pocket roughing routine

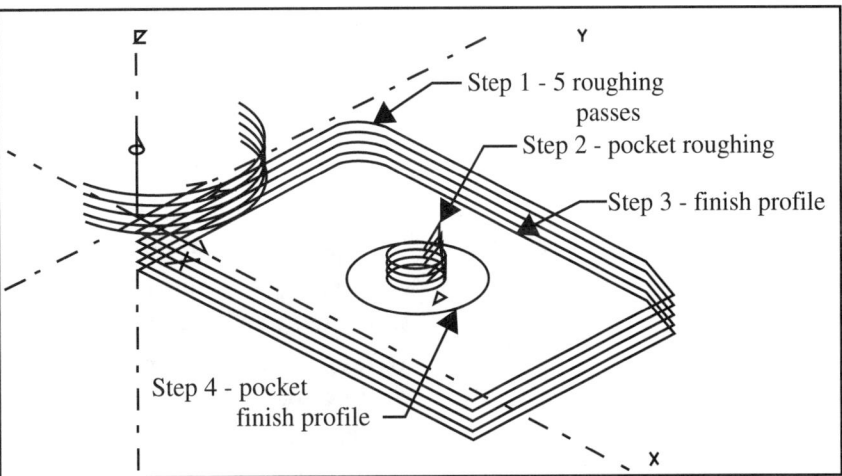

**FIGURE 4.22**
Open boundary error message

**FIGURE 4.23**
The input fields of the **Lead In/Out** control panel

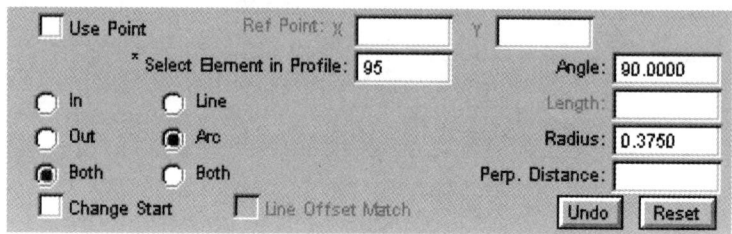

**Both** (lead in and lead out) was chosen for this example.

**Arc** was the type of lead in/lead out move that was chosen as opposed to a line lead in/lead out.

A value for **Angle** and **Radius** was chosen which would give the appropriate results for this example.

For **Select Element in Profile**, select the finish profile of the pocket. This should be the last input field that is filled in this control panel.

# Drilling the Holes

The next sequence of events will allow you to spot drill and then drill the holes in the process model. Before proceeding, obtain a full-scale, top view of the model.

1. Choose **Insert** from the workbench.
2. Choose the **After, Step Seq**, and **With Step** options from the tool list.

Fill in the input fields of your **Insert** control panel to match those in Figure 4.24.

This control panel is not unlike many of the others that you have already seen. There are a few things, however, that need to be mentioned.

**After Step** 4—Step 4 is the finish cut on the pocket. The spot drill (step 5) will come after the last endmill.

**Offset** is set to None—Remember, there is no offset on a drilling tool.

**Level** is set to 0.00—The cut will start at 0.00. The depth of the drill will be set in the **Hole** control panel.

**FIGURE 4.24**
The **Insert** control panel for step 5

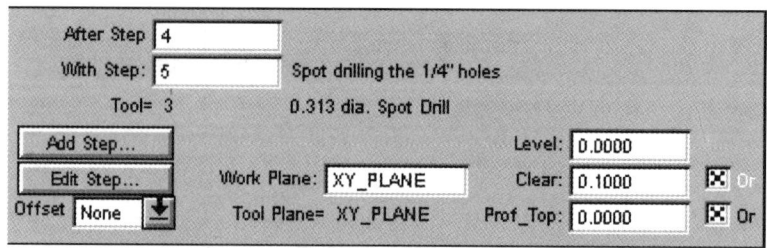

Chapter 4  SmartCAM Tutorial 3

3. Choose **Geometry** from the workbench.
4. Select **Hole** from the tool list.

Fill in the input fields of your **Hole** control panel to match those in Figure 4.25.

When this control panel is first opened you will notice many of the input fields are already filled in. Ignore these values and input .300″ into the **Spot Dia** field as the first step. The other fields will be updated accordingly.

**Hole Point** is the center point, according to your blueprint, of the first hole.

Upon entering the value for **Hole Point** SmartCAM will place the first spot-drilled hole in the proper location.

Continue by placing the drilled hole immediately after the spot drill.

1. Choose **Insert** from the workbench.
2. Choose the **After, Step Seq**, and **With Step** options from the tool list.
3. Enter "5" in the **After Step** input field.
4. Enter "6" in the **With Step** input field.

Accept the default for all other input fields. They will be the same as for the Spot Drill.

5. Choose **Geometry** from the workbench.
6. Choose the **Hole** tool from the tool list.

Fill in the input fields of your **Hole** control panel to match those in Figure 4.26.

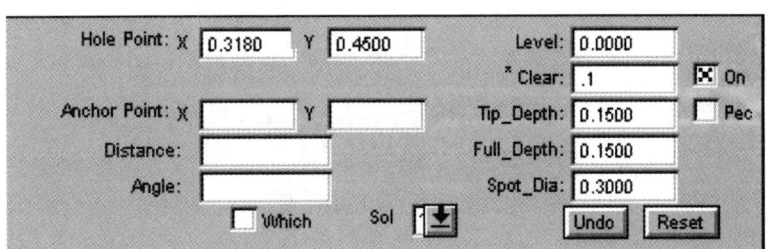

**FIGURE 4.25**
The **Hole** control panel

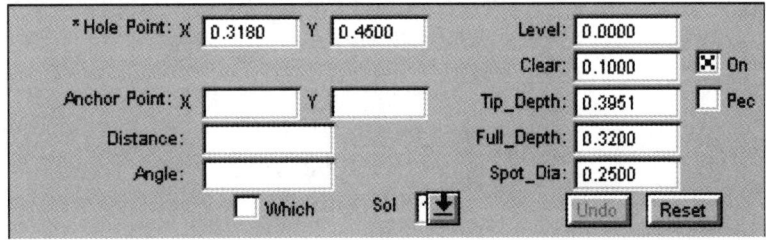

**FIGURE 4.26**
The **Hole** control panel

Again, when you open this control panel, many of the input fields will already contain values. Ignore these default values. Input .320 in the **Full Depth** input fields. All other values will be updated accordingly.

At this time, an explanation of these fields might help you to understand what you are doing a little better.

**Spot Dia** will calculate the "Z" depth of a tool when you specify a diameter the tool should cut, providing the information in the Job Planner is correct (drill point angle and drill diameter). Use this for spot drilling, chamfering, or countersinking.

**Full Depth** will calculate the tip depth of the tool when you enter the "Z" depth of the full diameter of the tool. Based on the blueprint, the 1/4" diameter hole is .32" deep. Enter this value into the **Full Depth** input field.

**Tip Depth** allows you to control the "Z" depth of the drill point.

**Hint:** *In SmartCAM, all of these values are positive!*

The **Clear** input field is the default value from the **Insert** control panel.

**Level** is the "Z" level at which the hole starts. It defaults to a value of 0.0 from the **Insert** control panel.

Since the drilled hole is located at the same coordinates as the spot drill, highlight the word **Hole Point** and pick the spot drill with the mouse. The Snap pick function must be on in order for this function to operate properly.

At this point, your process model should contain geometry representing a spot-drilled hole and a drilled hole in the lower left corner of the process model as per the blueprint.

Continue with the placement of the holes:

1. Choose **Group** from the workbench.
2. Choose the **New Group** button.
3. Select the **Element** option from the tool list.
4. With your mouse, select the last two elements of your database. This should be the spot-drilled hole and the drilled hole.
5. Choose **Edit** from the menu bar.
6. Select **Transform** from the menu.
7. Choose the **Move** tool.
8. Turn the **Copy** selection on.
9. Set **Copies** to 11.
10. Make sure the **Sort by Tools** option is on. This will automatically sequence your database so that all the spot-drilling is done before the drilling.
11. Choose **From 0**.
12. For the **To Point** input field, enter the following values:
    "X" **3.750 / 11**
    "Y" **0.00**
    "Z" **0.00**

At this point, you should have a row of spot-drilled and drilled holes across the bottom of the part.

13. Choose **Group** from the workbench.
14. Choose the **New Group** button.
15. Choose the **Step** option.
16. Choose step "5" and "6".
17. Choose **Transform** from the workbench. If it is not on the workbench it can be found under the Edit menu.
18. Choose the **Move** tool.
19. Turn the **Copy** selection on.
20. Set **Copies** to 1.
21. **Sort by Tools** should be turned on.
22. With your mouse, highlight the **From Point** input field.
23. Choose the **From 0** button.
24. For the **To Point** input field enter the following values:

    "X" **0.00**

    "Y" **2.00**

    "Z" **0.00**

At this point your process model is complete and should match Figure 4.27.

**FIGURE 4.27**
The completed process model

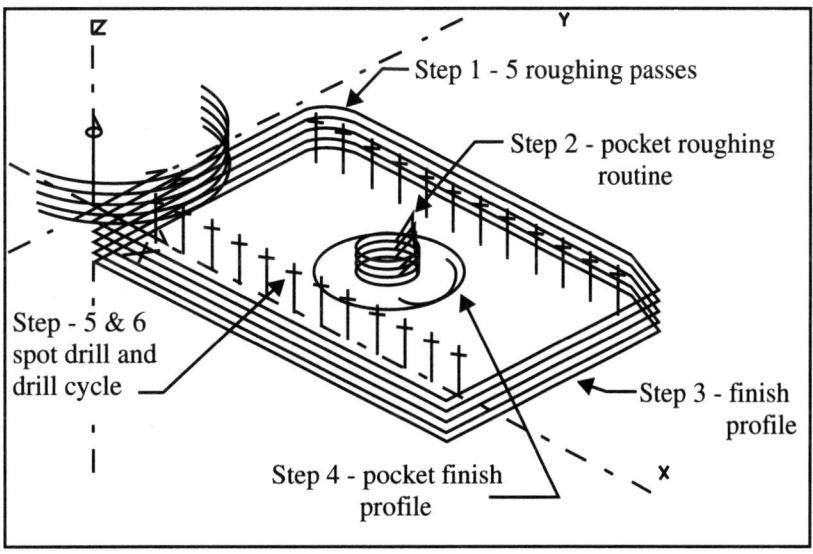

CHAPTER 5

# SmartCAM Tutorial 4

## Butterfly Flange

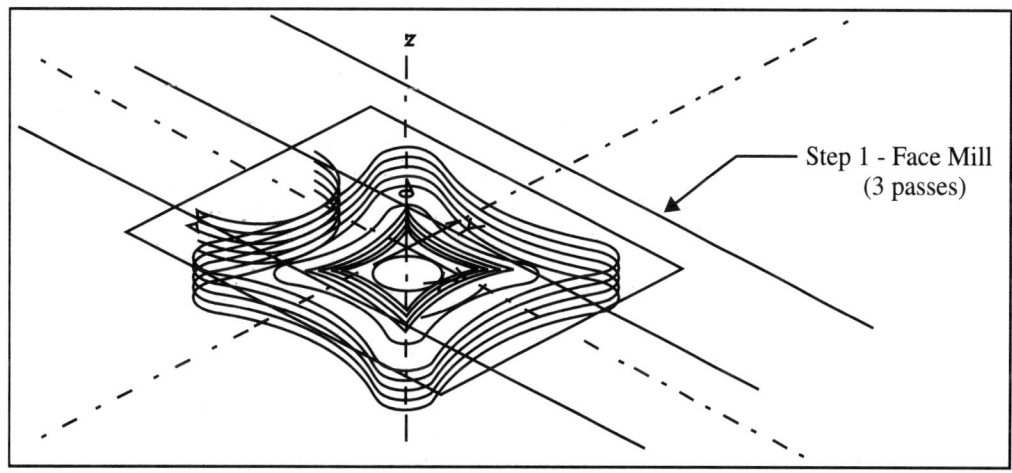

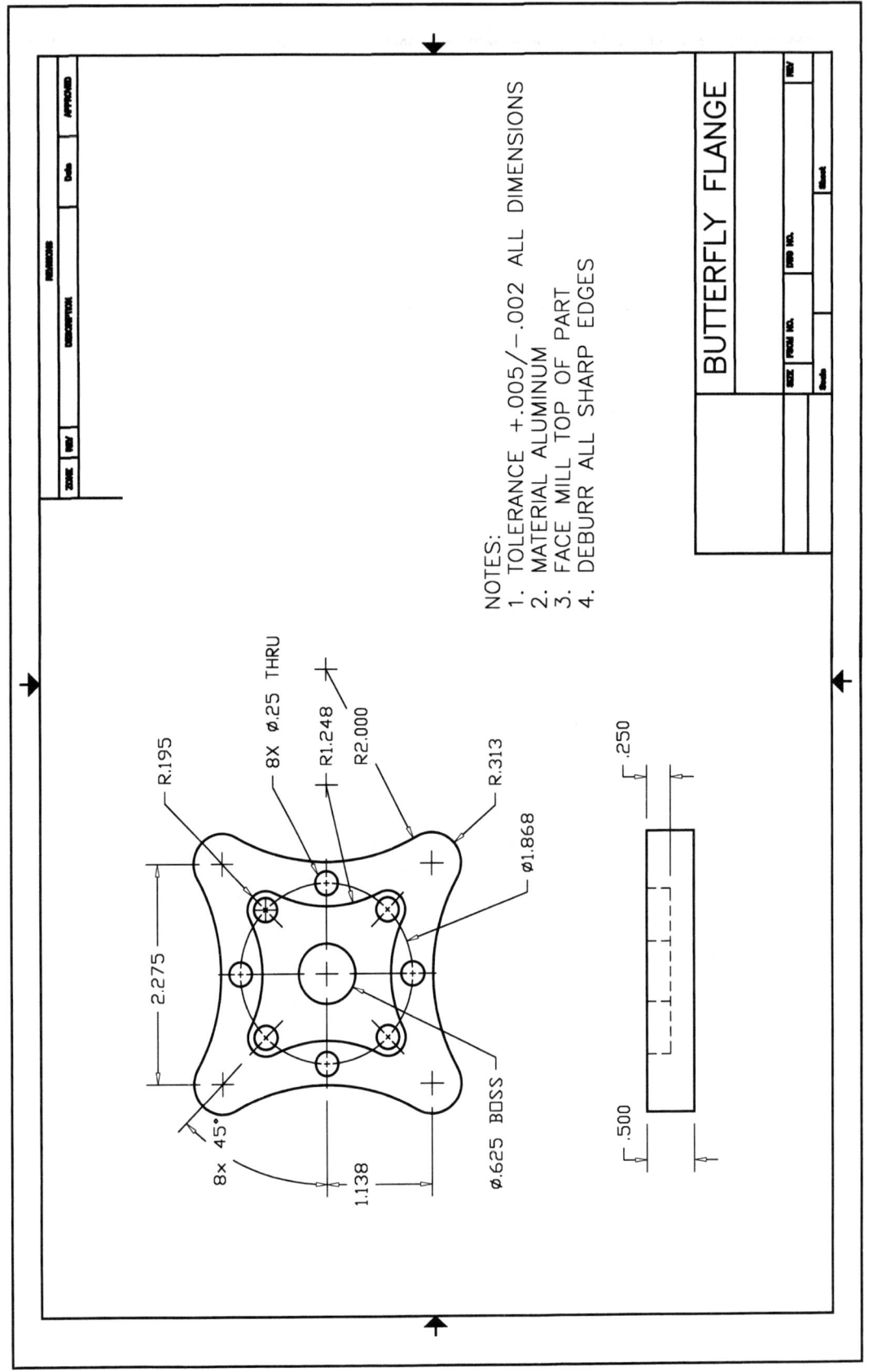

The blueprint for the Butterfly Flange Tutorial

Chapter 5  SmartCAM Tutorial 4

Upon completion of this chapter, you should be able to:

- Construct more complex geometry which consists of blended arcs.
- Construct pocket roughing routines which include islands.
- Correctly drill a bolt hole pattern which resides on different "Z" levels.
- Scale geometry.
- Face mill a workpiece to a specified thickness.
- Delete unwanted portions of geometry using the split and trim/extend functions.

# BUTTERFLY FLANGE

This tutorial will introduce you to several new procedures that are contained within SmartCAM. The previous two projects have consisted of regular geometric shapes. This project will consist of an irregular external contour and an irregular pocket. In addition, the pocket contains a boss, which is referred to as an "island." This will introduce you to SmartCAM's pocket roughing with island avoidance. The external geometry is symmetrical in shape, therefore, the applications of transforming your geometry can be well utilized in this project. Additionally, since the internal and external contours are identical in shape, this tutorial will also introduce you to the idea of scaling a contour. You will also deal with a circular bolt hole pattern that will introduce you to SmartCAM's rotate command.

As with the previous tutorials, the first process is to build a job operations file. The job operations file will consist of the following steps and tools:

**TABLE 5.1**

| Step # | Tool # | Type | Diameter | Speeds | Feeds |
|---|---|---|---|---|---|
| 1 | 1 | Face Mill | 3" | 600 SFPM | .005 IPT |
| 2 | 2 | End Mill (2 flute roughing) | 5/8" | 600 SFPM | .006 IPT |
| 3 | 3 | End Mill (4 flute finishing) | 5/8" | 650 SFPM | .005 IPT |
| 4 | 4 | End Mill (2 flute roughing) | 3/8" | 600 SFPM | .004 IPT |
| 5 | 5 | End Mill (4 flute finishing) | 3/8" | 650 SFPM | .003 IPT |
| 6 | 6 | Spot Drill | 3/8" (90 deg) | 600 SFPM | .006 IPR |
| 7 | 7 | Drill | 1/4" | 600 SFPM | .004 IPR |

# Constructing the Finish Profiles

When your Job Operations File is complete, you can then construct the external profile. Again, since your blueprint shows finish geometry, you must first build the external, finish toolpath. To build the external, finish toolpath, follow these steps:

1. Choose **Insert** from the workbench.
2. Choose the **After**, **Element Sequence**, and **With Step** options from the tool list.

    Fill in the input fields of your **Insert** control panel to match those of Figure 5.1.

3. Choose **Geometry** from the workbench.
4. Select the **Arc** tool from the tool list.
5. Fill in the input fields of your **Arc** control panel to match those of Figure 5.2.

Recall from the previous tutorial that SmartCAM will accept mathematical operations. If possible, always enter the values directly from the blueprint into the input field. For **Center Point** enter 2.275/2 for both the "X" and the "Y" input fields.

At this point you should have a full circle on your graphics work area at X 1.1375 and Y 1.1375.

In working with this tutorial, you will find that it will be easier for future operations if the center of the part is located at X 0, Y 0, rather than locating the corner of the part at X0, Y0. That's not to say that locating the corner of the part at X0, Y0 would not work, this is just a different way to approach this process model.

**FIGURE 5.1**
The input fields of the **Insert** control panel

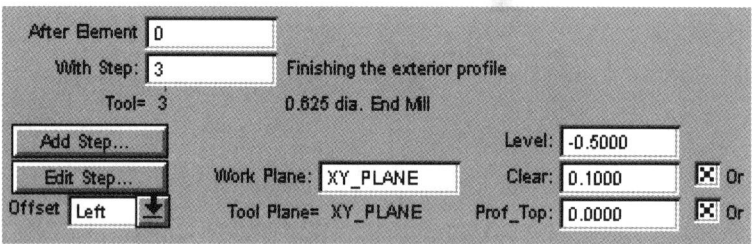

**FIGURE 5.2**
The input fields of the **Arc** control panel

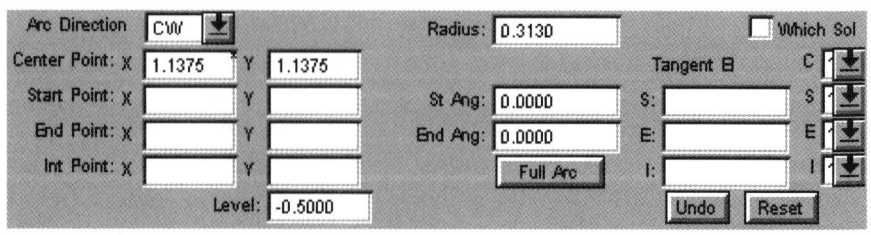

Chapter 5 SmartCAM Tutorial 4

To create the arc at the lower right corner of the process model, follow these steps:

1. Group the circle which was just created.
2. Choose **Transform** from the workbench. If it is not on the workbench, it can be found under the **Edit** menu.
3. Select the **Move** option from the tool list.
4. **Copy** is on and **Copies** is set to 1.
5. Choose **From 0**. When **Transform** is used to make multiple copies, the **From Point** and the **To Point** must be viewed as incremental values.
6. Choose **To Point**. The values must be X0.000, Y-2.275, Z0.000.

At the completion of the **To Point** input field the additional circle should be created.

Continue building your model by placing the 2″ arc between the two circles which you have already created.

1. Choose **Geometry** from the workbench.
2. Select the **Arc** tool from the tool list.
3. Set **Arc Direction** to CCW.
4. **Radius** should be set to 2.0″.
5. Verify that the **Level** is set to -.500.

At this time you will be introduced to another geometry building function of SmartCAM. On the right side of the **Arc** control panel you will see the heading **Tangent El**. Under that heading you will also see "S:" (which stands for start), "E:" (which stands for end), and "I:" (which stands for intermediate). Figure 5.3 shows this portion of the **Arc** control panel.

These functions will allow you to designate a tangent element start point, a tangent element end point, and if needed, a tangent element intermediate point.

1. With your mouse, highlight the **S** input field.
2. Pick the lower, right quadrant of the upper circle.

**FIGURE 5.3**
The Tangent Element function of the **Arc** control panel

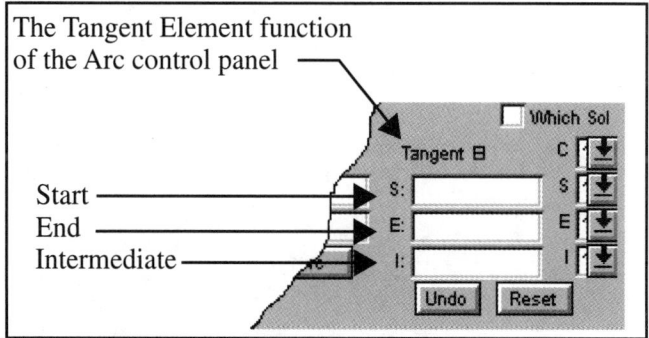

**FIGURE 5.4**
The tangent element start and end points

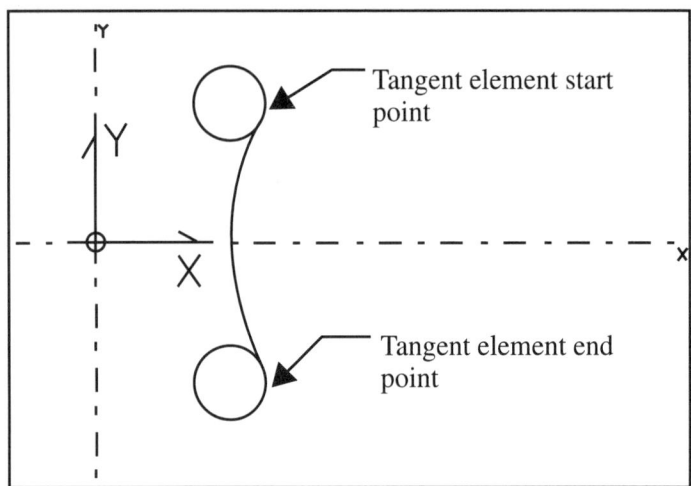

3. With your mouse, highlight the **E** input field.
4. Pick the upper, right quadrant of the lower circle.

SmartCAM will place the appropriate arc between the two circles. Figure 5.4 shows the correct mouse pick points as well as the intended results.

Continue building your process model by grouping the entire model.

1. Choose **Group** from the workbench.
2. Choose the **Select All** button.

At this point, instead of creating the rest of the geometry from scratch, simply utilize your existing geometry to generate the rest of the profile.

1. Choose **Transform** from the workbench. If it is not on the workbench, it can be found under the **Edit** menu.
2. Choose the **Mirror Image** option from the tool list.

   (If, by chance the Mirror Image option is grayed out, verify that an active group exists.)
3. Fill in the input fields of your **Mirror Image** control panel to match those of Figure 5.5.

   For the purposes of this example, the **2D** option should be on.

   **First Point** should be set to an arbitrary point on the Y axis.

**FIGURE 5.5**
The **Mirror Image** control panel

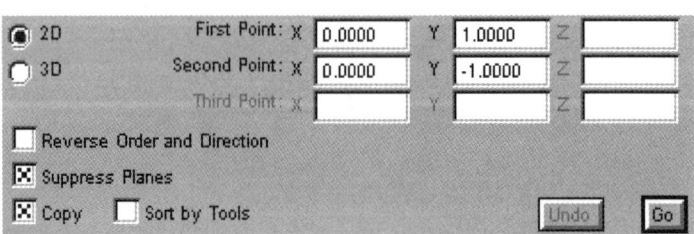

**Second Point** should be set to a different point on the Y axis.

**Suppress Planes** should be turned on. With this option turned off, SmartCAM will create a new work plane on which the new geometry is placed. This may cause trouble when the code is generated.

**Copy** should be turned on in order to mirror an additional copy. With **Copy** turned off the geometry will simply be moved to the other side of the Y axis.

When these fields are properly filled in, click on the **Go** button.

At this point of the process, your model should match Figure 5.6.

In order to complete the external profile, you need to rotate the two 2" arcs 90 degrees.

1. Choose **Group** from the workbench.
2. Select the **New Group** button to deactivate the existing group.
3. Choose the **Element** option and choose the two 2" arcs.
4. Choose **Transform** from the workbench.
5. Select the **Rotate** option from the tool list.
6. Fill in the input fields of your **Rotate** control panel to match those of Figure 5.7.

As with the **Mirror Image** input field, you are doing a **2D** rotation.

**Pivot Axis Point** is X0,Y0 or the center of the model.

**FIGURE 5.6**
The result of the **Mirror Image** command

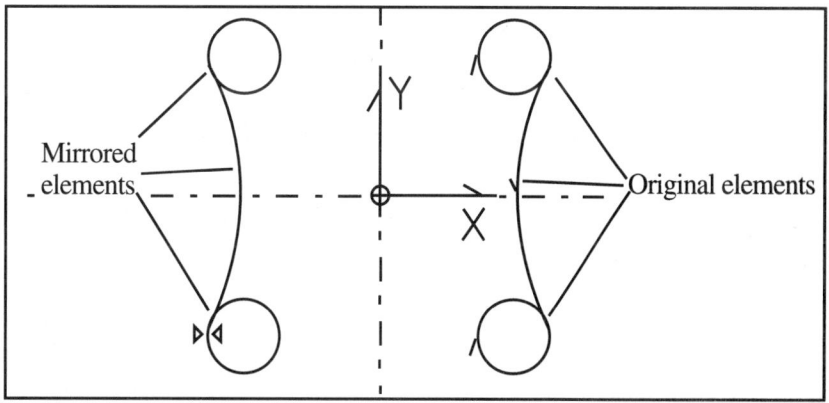

**FIGURE 5.7**
The input fields of the **Rotate** control panel

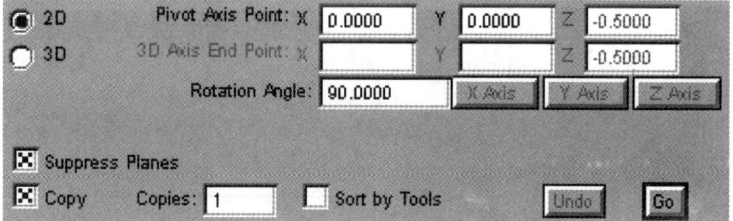

**Rotation Angle** is set to 90 degrees.

**Suppress Planes** is again turned on.

**Copy** is turned on.

**Copies** is set to 1.

When these fields are properly filled in, click on the **Go** button.

At this point, your process model should match Figure 5.8.

Continue building the model by trimming out the unwanted portion of the .313″ radius corners.

1. Choose **Edit** from the menu bar.
2. Choose the **Geo Edit** toolbox.
3. Choose the **Trim/Extend** option.

**Remember:** *From the previous trimming functions, to pick the arc where you want to* keep *it, not where you want to* remove *it.*

Getting the intended results when using the **Trim/Extend** function may prove to be a little difficult at times. For the function to work properly, you must keep track of the arc direction, the start and end points of the arc, and the location at which the arc is selected. This can be a little tedious. However, there is an easier way to deal with this issue. Prior to choosing the elements to trim, turn on the **Which Segments** switch in the **Trim/Extend** control panel. This will, upon selection of the elements to trim, open a dialog box with **Previous**, **Next**, **Accept**, and **Cancel** switches. The **Previous** and **Next** selector switches will allow you to view all possible solutions to your trim function. When SmartCAM shows the correct solution, simply pick the **Accept** button.

Occasionally you will need to recreate a portion of the arc that, due to the execution of the trim function, failed to reconnect entirely. Simply use the **Trim/Extend** function to reconnect all geometry.

Upon completion of the trimming and extending operation, your process model should match Figure 5.9.

**FIGURE 5.8**
The result of the **Rotate** command

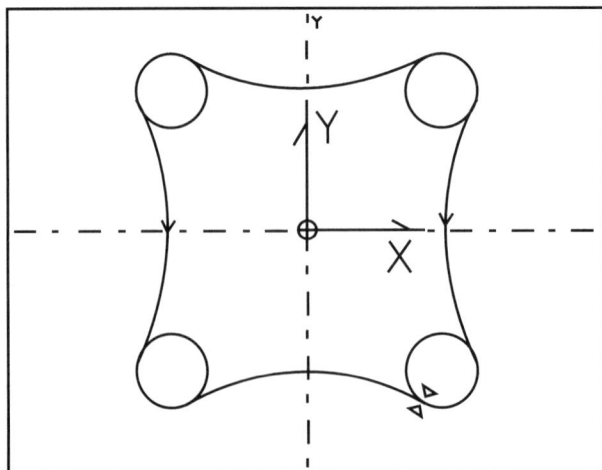

**FIGURE 5.9**
Elements have been grouped to indicate the element direction

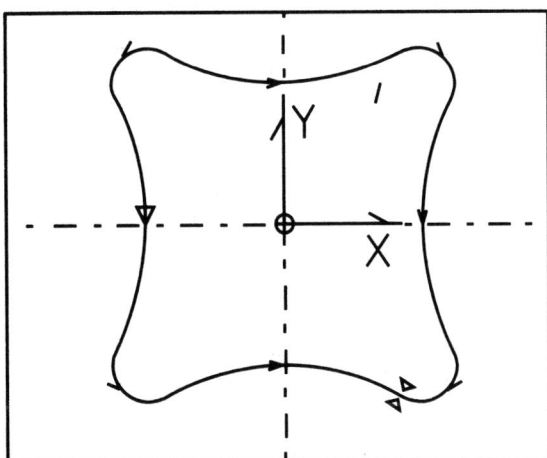

If you will notice from Figure 5.9 and also your own process model, the geometry is out of sequence and the direction of the elements is both clockwise and counterclockwise. Before you can do anything else, it is important that your database is properly sequenced and all elements progress in a clockwise manner. In order to accomplish this task, follow these steps:

1. Choose **Group** from the workbench.
2. Choose the **Select All** button.
3. Choose **Edit** from the menu bar.
4. Choose the **Order Path** toolbox from the menu.
5. Select the **Chain** tool from the tool list.

**Chain** allows you to alter two properties related to your external geometry. First, the **Chain** tool will properly sequence the database. Second, if you choose the **Group Chain** option, the active group of elements will progress in the direction that is the same as the first element of the database. If, however, you choose **Select an Element in profile**, all of your elements will progress in the same direction as the element that you select. Additionally, this element becomes the first element of your database.

6. With your mouse, select the **Group Chain** button.

At this time, all elements should progress around the profile in a clockwise manner. This will be evident by the direction of the arrowheads representing the active group.

If at any time the geometry fails to chain properly, analyze the "Z" depth of the geometry. This can be done by following these steps:

1. Choose **Utility** from the menu bar.
2. Select the **Element Data** option.
3. Select each element from the graphics work area.

All "Z" depths need to be the same to create a proper profile. The top view of the model may be deceiving since depth cannot be viewed.

## Constructing the Pocket Finish Profile

As indicated by the blueprint, the interior profile is nothing more than a scaled-down version of the exterior profile. This will provide you an opportunity to work with SmartCAM's **Scale** function.

Make sure that all geometry is in sequential order and grouped.

1. Choose **Edit** from the menu bar.
2. Choose the **Transform** toolbox from the selections.
3. Select the **Scale** tool from the tool list.

Upon opening the **Scale** control panel, you will notice input fields for the "X," "Y," and "Z" axes as shown in Figure 5.10. The value for these input fields is nothing more than a ratio by which SmartCAM will either increase or decrease the size of the grouped geometry. It is important to understand that the scale factor of 1.0 will not change the size of the profile. Positive numbers less than 1.0 will shrink the geometry and numbers greater than 1.0 will enlarge the geometry. The question is, what value will give you the proper size profile? If you will notice, your blueprint specifies a 2.000" radius and a 1.248" radius. Since your desired profile is to be smaller, simply enter 1.248/2.000 into the **X Factor** and **Y Factor** input fields. The actual value will be .624", however it is best to enter the values directly from your print.

The value for **Z Factor** should be 1.0. You do not want to scale this profile in the "Z" axis.

The value for **Reference Point** should be X 0.000, Y 0.000, and Z -.500. When these input fields are correct, click the **Go** button.

The first thing you notice is that the original profile disappeared. That is because there is no copy function with **Scale**. Choose the **Undo** button and the geometry will return to its original size.

In order to accomplish this task, you must first create a duplicate copy of your external profile.

1. Choose **Transform** from the workbench. (If it is not on the workbench, it is found under the **Edit** menu.)
2. Choose the **Move** tool from the tool list.

**FIGURE 5.10**
The **Scale** control panel

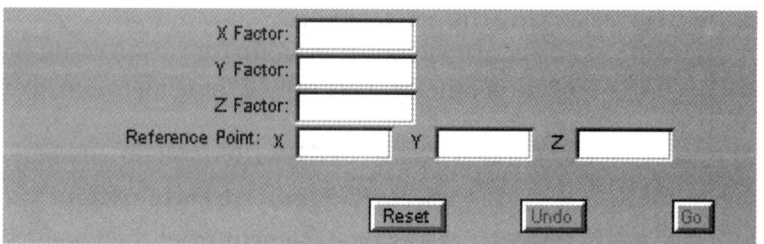

# Chapter 5  SmartCAM Tutorial 4

3. Turn **Copy** on.
4. Set **Copies** to 1.
5. With your mouse, choose the **From 0** button.
6. For **To Point**, enter X 0.00, Y 0.00, Z -.0001.

Notice the database reflects sixteen elements, which is the quantity of two profiles. Now, return to scale and fill in the input fields as you did previously. When the input fields are complete, select the **Go** button.

At this time, your process model should match Figure 5.11.

At this point, analyzing the pocket profile visually and with **Element Data**, found under the **Utility** menu, will reveal several problems with the geometry. Currently the "Z" depth of the pocket profile is incorrect, the step is incorrect, the tool offset is incorrect, and the geometry is progressing in the wrong direction.

To correct these problems, follow these steps:

1. Choose **Edit** from the menu bar.
2. Select the **Order Path** toolbox from the menu.
3. Choose the **Reverse Order** tool from the tool list.
4. Inside the **Reverse Order** control panel, choose the **Direction Only** option.
5. With your mouse, select the **Group Reverse** button as shown in Figure 5.12.

**FIGURE 5.11**
Original and scaled profiles

**FIGURE 5.12**
The **Group Reverse** button

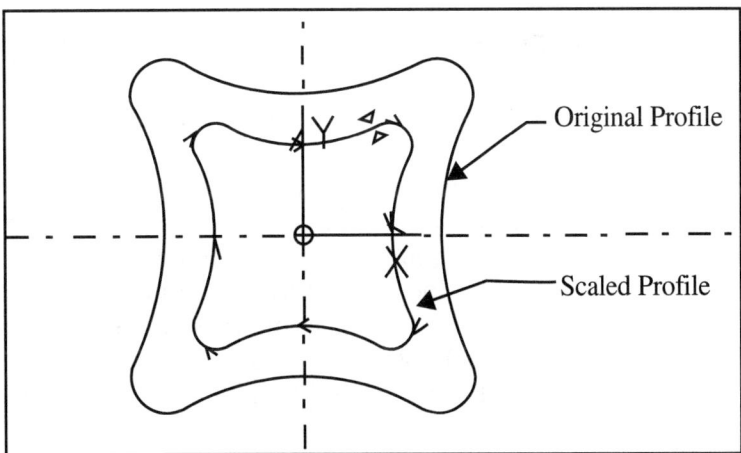

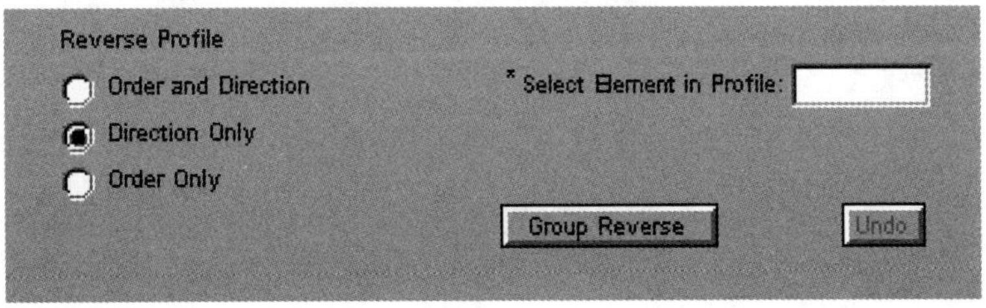

With the activation of the Group Reverse button, the geometry should progress in the required counterclockwise manner.

6. Again, choose **Edit** from the menu bar.
7. Select the **Property Change** option from the menu.
8. Select the **Toolpath** option.
9. Fill in the **Toolpath Property Change** dialogue box to match Figure 5.13.

The **Toolpath Property Change** control panel is very similar to the **Insert** control panel in that the input fields are the same. The **Toolpath Property Change**, however, will only allow you to update the properties of *existing* geometry.

To reevaluate the elements that make up the pocket profile, follow these steps:

1. Choose **Utility** from the menu bar.
2. Select the **Element Data** option of the menu list.
3. Verify that the pocket profile has a "Z" depth of -.250".
4. Verify the step number is 5.
5. Verify the offset of the elements is left.
6. Group the profiles and verify that the geometry is progressing in a counterclockwise manner.

## Constructing the "Island"

In order to complete the pocket profile, you must now construct the island in the center of the pocket. This is an opportunity to identify the purpose of **Match Element** that is found in the tool list of the **Insert** toolbox. Before you proceed, once again verify the element data of the pocket profile. The profile top should be set to 0.00 and the "Z" level should be set to -.250". SmartCAM looks at the "Z" level and the profile top of an element to establish a thickness. It is critical that these values are correct—otherwise, the roughing routine will not avoid the island. If any values are incorrect, simply group the incorrect elements and change them with **Toolpath Property Change**.

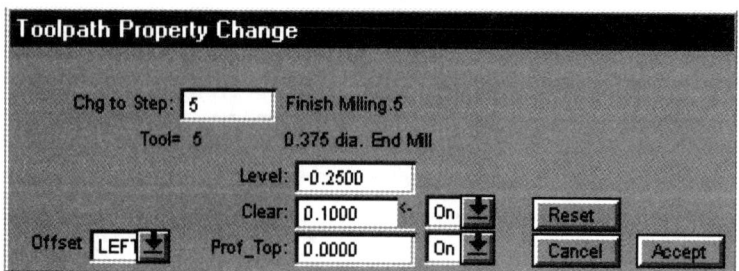

**FIGURE 5.13**
The **Toolpath Property Change** control panel

1. Choose **Insert** from the workbench.
2. Choose the **After** and the **Step Sequence** options from the tool list.
3. Turn **Match Element** on. Remember, an "X" in the box indicates the on condition. With **Match Element** on, the **Insert** control panel input fields will default to the properties of the step sequence you choose.
4. To complete the operation, simply pick the pocket profile.

All input fields of the Insert control panel should now match the properties of the pocket profile. To construct the geometry, follow these steps:

1. Choose **Geometry** from the workbench. If by chance **Geometry** is not on the workbench, it can be found under the **Create** menu of the menu bar.
2. Select the **Arc** tool from the tool list.
3. Create a clockwise, full arc, with a radius of .625/2. The center is located at X 0.00 and Y 0.00.

At this point, your process model should match Figure 5.14.

## Roughing the External Geometry

The roughing routine for the exterior profile is exactly the same as in the previous tutorial. However, before you begin, you must add a lead in/lead out move to your geometry. The lead in/lead out elements need to be placed in an area where there is maximum clearance. Therefore, it should be placed on the midpoint of the left side of the model.

Recall that a lead in move will automatically be placed at the beginning of the profile. You must modify your external profile so the first element of the profile is at the midpoint of the left arc of the model.

1. Choose **Geo Edit** from the workbench. If it is not on the workbench, it may be found under the **Edit** menu of the menu bar.

**FIGURE 5.14**
The external profile, the island, and the pocket profile

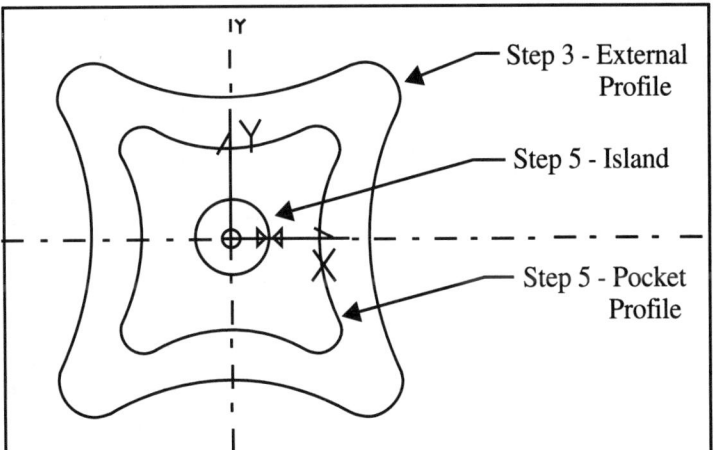

**FIGURE 5.15**
The 2″ arc split at 50 percent of its length

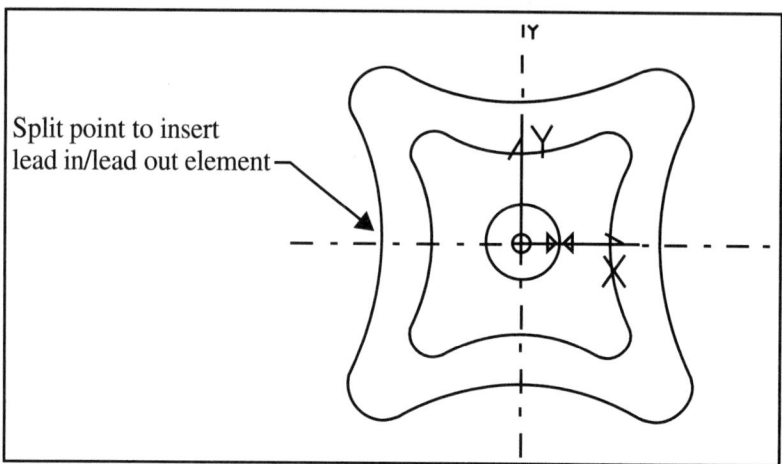

2. Select the **Split** tool from the tool list.
3. Choose the **Element Division** option of the **Split** control panel.
4. Split the 2.0″ arc at 50 percent of the length. Refer to Figure 5.15 for further clarification.
5. Choose **Geo Edit** from the workbench.
6. Choose the **Lead In/Out** option from the tool list.
7. Inside the **Lead In/Out** control panel:
   a. Choose **Both**.
   b. Choose **Arc**.
   c. Turn **Change Start** on.
   d. Enter an **Angle** and a **Radius** that you feel is appropriate.
   e. Choose **Select Element in Profile** and pick the left, 2″ radius above the midpoint.

The lead in/lead out geometry should be added to the profile as shown in Figure 5.16.

**FIGURE 5.16**
The lead in/lead out geometry

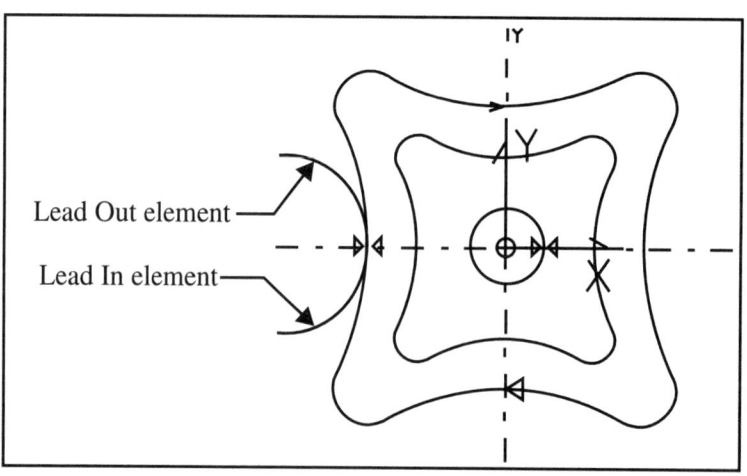

# Chapter 5  SmartCAM Tutorial 4

After you have completed the lead in/lead out geometry, the next step is to create a roughing profile from the finish geometry.

1. Choose **Group** from the workbench.
2. With your mouse, select the **New Group** button to degroup any active groups.
3. Choose the **Profile** option from the tool list.
4. With your mouse, select the external profile.
5. From the workbench, choose **Insert**.
6. Select the **Before**, **Step Sequence**, and **With Step** tools from the tool list.

**Caution:** *Make sure Match Element is off.*

Fill in the input fields of your **Insert** control panel to match those of Figure 5.17.

1. Choose **Geometry** from the workbench.
2. Select the **Wall Offset** tool from the tool list.
    a. Set **Wall Side** to Left.
    b. Enter a value of .010" into **Distance**.
    c. **Wall Repeats** should be set to 1.
3. Choose the **Group Wall** button.

You should now have a roughing profile at Z -.100 as shown in Figure 5.18.

**FIGURE 5.17**
The input fields of the **Insert** control panel

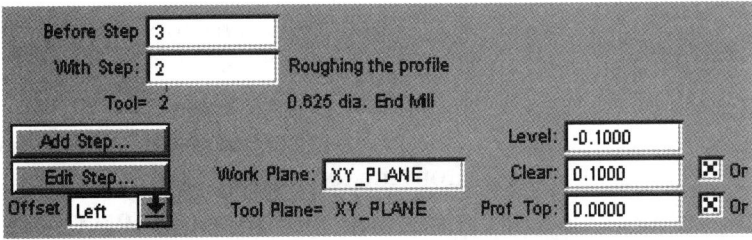

**FIGURE 5.18**
The result of the wall offset function

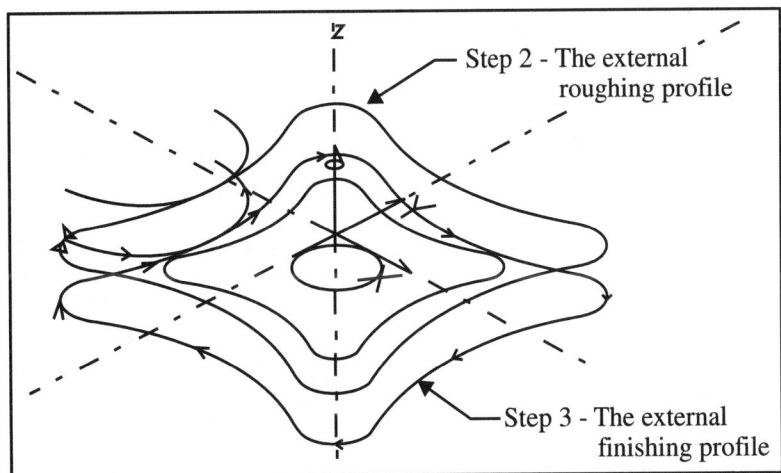

Continue the creation of the roughing profiles by grouping Step 2, following these steps:

1. Choose **Group** from the workbench.
2. With your mouse, select the **New Group** button.
3. Choose the **Step** option of grouping.
4. Choose Step 2, the roughing step.
5. Choose the **Edit** menu from the menu bar.
6. Select the **Transform** toolbox.
7. Choose the **Move** tool from the tool list.
8. Turn **Copy** on.
9. Set **Copies** to 4.
10. For the **From Point** input field, select the **From 0** button.
11. For the **To Point** input fields, enter X 0.000, Y 0.000, and Z -.1000.

Upon completion of step 11, your process model should match Figure 5.19.

# Creating the Pocket Roughing Routine

In order to rough the material out of the pocket, it is necessary to obtain a full-scale, top view of the model. Do so before proceeding.

1. Choose **Group** from the workbench.
2. With your mouse, select the **New Group** button.
3. Choose the **Element** tool and pick the .3125″ radius circle which represents the island.
4. Choose **Insert** from the workbench.
5. Choose the **Before**, **Step Sequence**, and **With Step** options from the tool list.

Fill in the input fields of your **Insert** control panel to match those of Figure 5.20.

**FIGURE 5.19**
The completion of the external roughing profiles

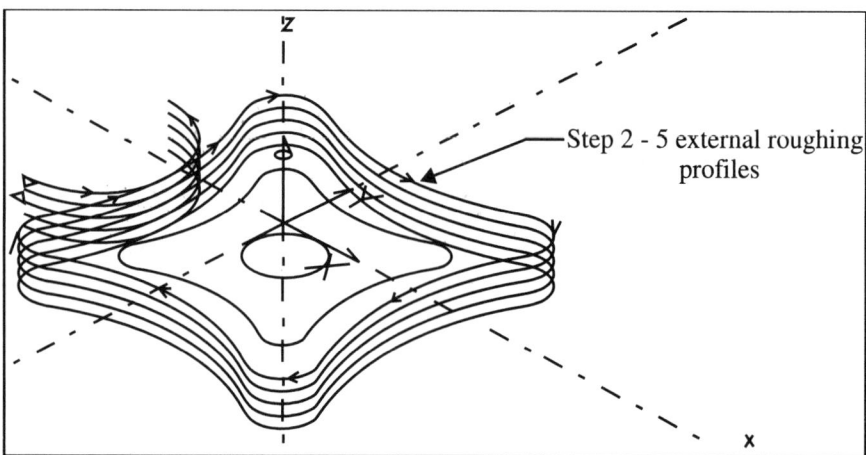

**FIGURE 5.20**
The input fields of the **Insert** control panel

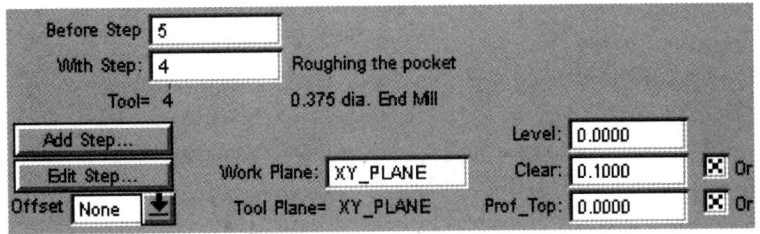

1. Choose **Process** from the menu bar.
2. Select the **Rough** tool box.
3. Choose the **Pocket** option.

Fill out the input fields of your **Pocket** control panel to match those of Figure 5.21.

**Pocket** works best for this example when set to Spiral. However, the other options will work. Since you have chosen Spiral as your pocket toolpath, you may want to take a minute and check the settings in the **Spiral Parameters** dialog box. These settings will affect the outcome of your toolpath.

**Final Island Pass** should be turned on.

**Group Island** is turned on. With the circle grouped, the roughing passes will avoid this area.

**Hint:** *For this to work properly, both the circle and the pocket periphery must have a profile top of zero and a level of -.250".*

**Climb Cut** is turned on.

**Outside Boundary** will be any element that makes up the pocket profile. After highlighting this input field, choose the pocket profile with your mouse. This input field was intentionally left blank in this example. Your input may vary depending upon the number of elements in your database list.

**Width of Cut** will default to half the tool diameter. However, smaller step over values, such as the one entered, seem to give better results.

**Finish Allowance** is the amount of material you want to leave on both the pocket profile and the island profile to be removed by the finish tool.

**FIGURE 5.21**
The input fields of the **Pocket** control panel

**FIGURE 5.22**
The completed pocket roughing routine

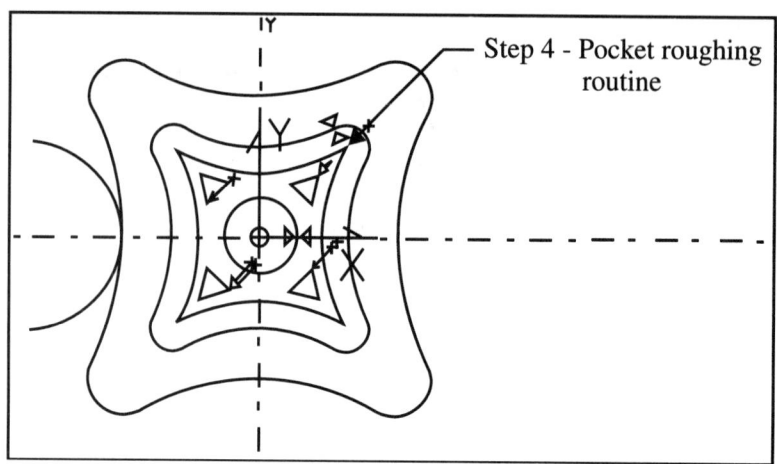

**First Pass Level** is the depth of cut for the first pass. This must be a negative value.

**Depth of Cut** is the depth for all successive passes. This must be a positive value.

**Final Pass Level** is the final depth of the pocket.

**Floor Allowance** is set to 0.00. No material will be left on the floor of the pocket for the finish tool to remove.

**User Start Point** must be left blank. Do not set it to 0.00. This is an absolute value that is outside of the pocket.

When these input fields are correct, choose **Go**. Your process model should now match Figure 5.22.

After the completion of the pocket roughing routine, lead in/lead out geometry can be placed on both the island and the pocket profile. Refer to the previous examples and Figure 5.23 for the proper construction and placement of the lead in/lead out geometry.

**FIGURE 5.23**
The proper construction and placement of the lead in/lead out geometry

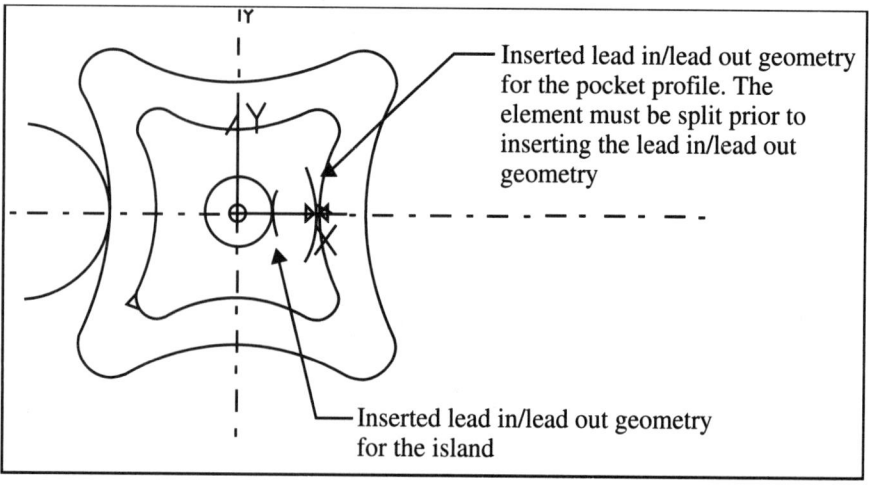

# Constructing the Bolt Hole Circle

You are now ready to spot drill and drill the bolt hole circle.

1. Choose **Insert** from the workbench.
2. Select the **After**, **Step Sequence**, and **With Step** options from the tool list.
3. Fill in the input fields of your **Insert** control panel to match Figure 5.24.

To place the spot drill in the proper location, follow these steps:

1. Choose **Geometry** from the workbench.
2. Choose the **Hole** tool.
3. Input a **Spot Dia** of .250".
4. Input a **Hole Point** of X 1.868/2 and Y 0.00.

A point, representing a hole, should appear in the graphics work area.

When you have completed the spot drill control panel, follow these steps:

1. Choose **Insert** from the workbench.
2. Select the **After**, **Step Sequence**, and **With Step** options from the tool list.
3. Enter **After Step** 6, **With Step** 7.
4. Accept the defaults for all other input fields (they should be the same as the previous illustration).

To place the drill in the proper location, follow these steps:

1. Choose **Geometry** from the workbench.
2. Again, choose the **Hole** tool from the tool list.
3. Input a **Full Depth** of .500".
4. With your mouse, select the word **Hole Point** from within the control panel.
5. Pick the existing spot drill location from your process model working area.

**FIGURE 5.24**
The input fields of the **Insert** control panel

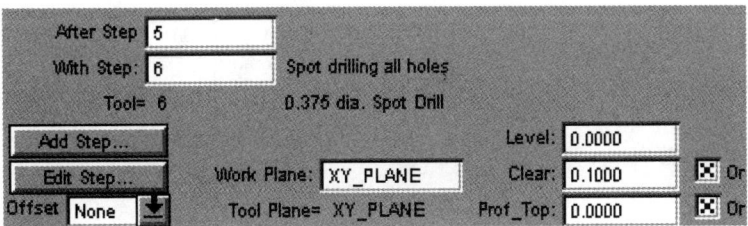

To pick the spot drill from the graphics working area, snap must be on and the Edit filter dialog box must be set to recognize holes.

Geometry representing the spot drilled and the drilled holes should now be visible in the graphics work area. Your database should also list the additional geometry.

To create the geometry for the other seven holes, follow these steps:

1. Choose **Group** from the workbench.
2. Select the **New Group** button.
3. Select the **Step** tool from the tool list.
4. From the database list of the available steps, choose step 6 and step 7.
5. Choose **Edit** from the menu bar.
6. Choose the **Transform** toolbox.
7. Select the **Rotate** tool from the tool list.
8. Fill in the input fields of your **Rotate** control panel to match those of Figure 5.25.

    **2D** rotation is appropriate for the XY plane.

    **Suppress Planes** should be on.

    **Copy** should be on.

    **Copies** should be set to 7 (8 minus the original).

    **Pivot Axis Point** is the XY center of the bolt circle.

    **Rotation Angle** is set to 360/8.

    **Sort by Tools** is set to on. This will prevent excessive tool changes.

9. Select the **Go** button.

At this point, all of the spot drill and drill tools are in their proper places. However, all of the holes have the same starting Z value of 0.00. This is fine for the holes outside the pocket. But the holes in the pocket would have to feed through .250″ of air before they came in contact with material. This is not desirable for today's production standards.

To solve this problem you need to move the four holes in the pocket to the proper Z level of -.250″.

In order to view the hole geometry a little more clearly, mask all steps except steps 6 and 7.

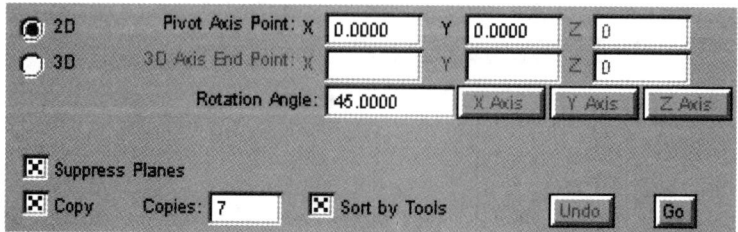

**FIGURE 5.25**
The input fields of the **Rotate** control panel

1. Choose **Utility** from the menu bar.
2. Select the **Show/Mask** option.
3. Hide steps 1 through 5.
4. When the database list shows an "H" beside steps 1 through 5, choose **Accept**.
5. Now, obtain a full-scale top view of the process model.
6. Choose **Group** from the workbench.
7. With your mouse, select the **New Group** button.
8. Select the **Box** option from the tool list, and group—by a box—the four holes that are in the pocket.
9. Choose **Edit** from the menu bar.
10. Select the **Property Change** option.
11. Select the **Holes/Points** option.
12. Input a new **Level** of -.250".

At this point, your process model, with tools 1-5 masked, should match Figure 5.26. The workplane indicator, as well as the axis indicators have been turned off.

Upon visual inspection, you will notice that the holes in the pocket are deeper than the other four. This is because the full depth that was input into the **Hole** control panel is an incremental distance from the Z level. The Z level of the holes on the outside of the pocket is 0.00, therefore our .575" hole depth is the proper depth. However, the Z level of the holes inside the pocket is -.250". Our holes inside the pocket are therefore .250" too deep.

To correct this problem, make sure that only the four drilled holes within the pocket are in the active group. Since both the spot drills and the drills in the pocket are in the active group, we must remove the spot drills. To remove the spot drills, follow these steps:

1. Choose **Group** from the workbench.
2. Make sure the **Remove** option is turned on.

**FIGURE 5.26**
The completion of the 5 drilled holes

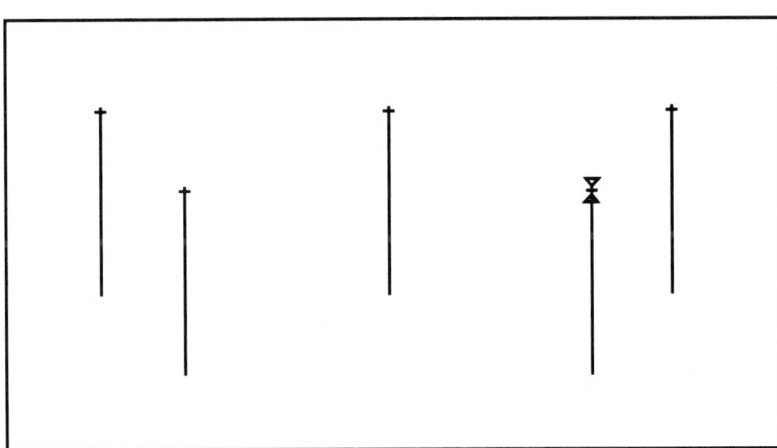

3. Choose **Step** and choose step 6. This will remove the spot drill from the active group.
4. Choose **Edit** from the menu bar.
5. Select **Property Change** from the menu list.
6. Select the **Holes/Points** option.
7. Input the value of .250″ into the **Full Depth** input field.

   .250″ is the material thickness from the bottom of the pocket to the bottom of the part.
8. Choose **Accept**.

At this point all holes should be the same depth as reflected in the front view.

## Face Milling the Workpiece

The last step in this tutorial is placing the face mill geometry on the model. Face milling is a roughing cycle, and a roughing cycle requires a closed boundary. Currently, you have no closed boundary representing your workpiece. You must, therefore, draw a rectangular box around your model to create a closed profile. This is best done on a layer.

All geometry is currently hidden from the previous exercise. You must show all geometry:

1. Choose **Utility** from the menu bar.
2. Choose the **Show/Mask** option.
3. Inside the **Show/Mask** control panel, turn the **Show** option on.
4. Select the **All** button.
5. Choose **Insert** from the workbench.
6. Select the **After**, **Step Sequence**, and **On Layer** options from the tool list.
7. Inside the **Insert** control panel, choose **After Step** 7, **On Layer** 1, and **Level** 0.00.
8. Choose **Geometry** from the workbench.
9. Choose the **Line** tool.
10. Build a rectangular box which completely encloses the process model profile, as shown in Figure 5.27. This will be your outside boundary for the face mill roughing cycle.

When you have completed the rectangular box, follow these steps.

1. Choose **Insert** from the workbench.
2. Choose the **Before**, **Step Sequence**, and the **With Step** options from the tool list.

Fill in the input fields of your **Insert** control panel to match those of Figure 5.28.

**FIGURE 5.27**
The outside boundary for the face mill roughing cycle

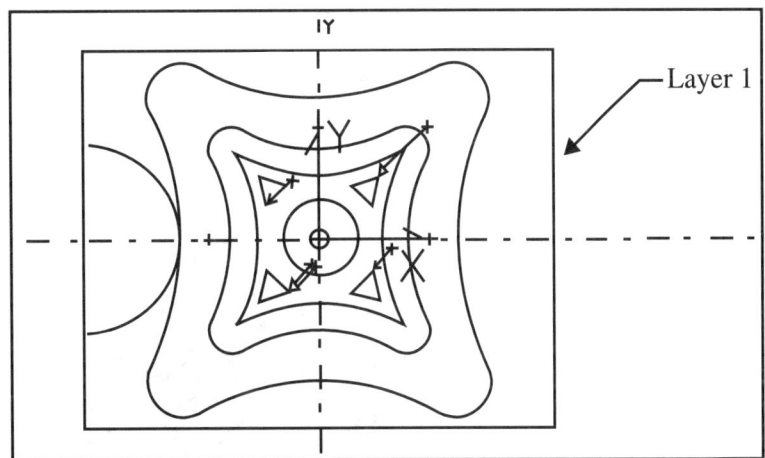

**FIGURE 5.28**
The input fields of the **Insert** control panel

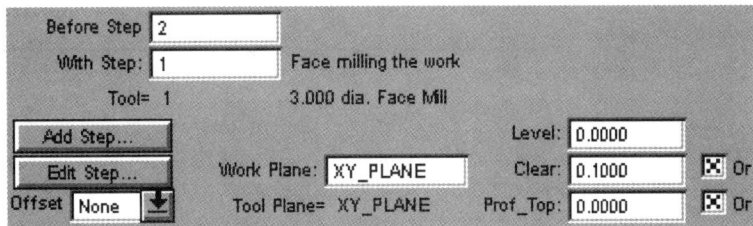

1. Choose **Process** from the menu bar.
2. Choose the **Rough** tool box.
3. Select the **Face** tool.
4. Fill in the input fields of your **Face** control panel to match those of Figure 5.29.

    **Face** is set to **Linear**. The other options will work equally as well.

    For **Outside Boundary** pick the boundary that you constructed on a layer.

    For the purposes of this example, all depths of cut values were set to 0.000. In order to achieve a depth of cut on the face mill, the G54 "Z" value of the machine tool will be set to .050" after touching off all tools.

    Accept the defaults for all other input values.

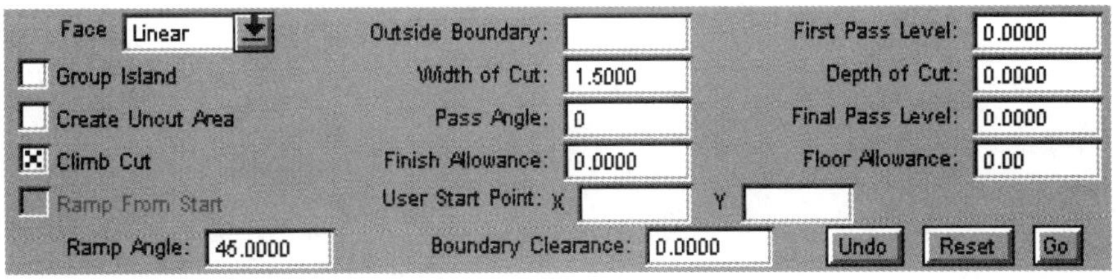

**FIGURE 5.29**
The input fields of the **Face** control panel

5. Select the **Go** button.

This concludes the tutorial on the Butterfly Flange. Your completed process model should match Figure 5.30.

**FIGURE 5.30**
The completed Butterfly Flange process model

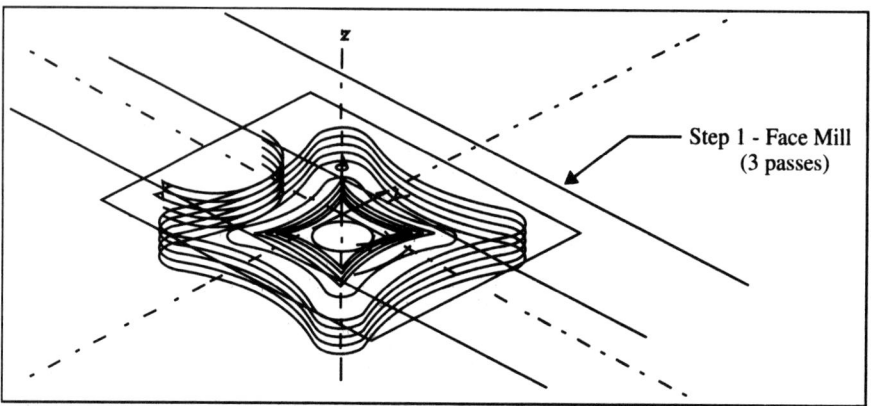

# CHAPTER 6

# SmartCAM Tutorial 5

## Four-Hole Frame

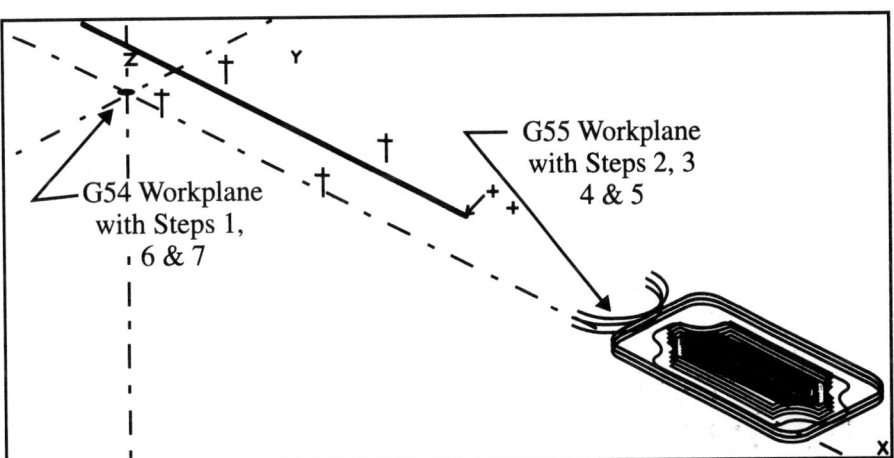

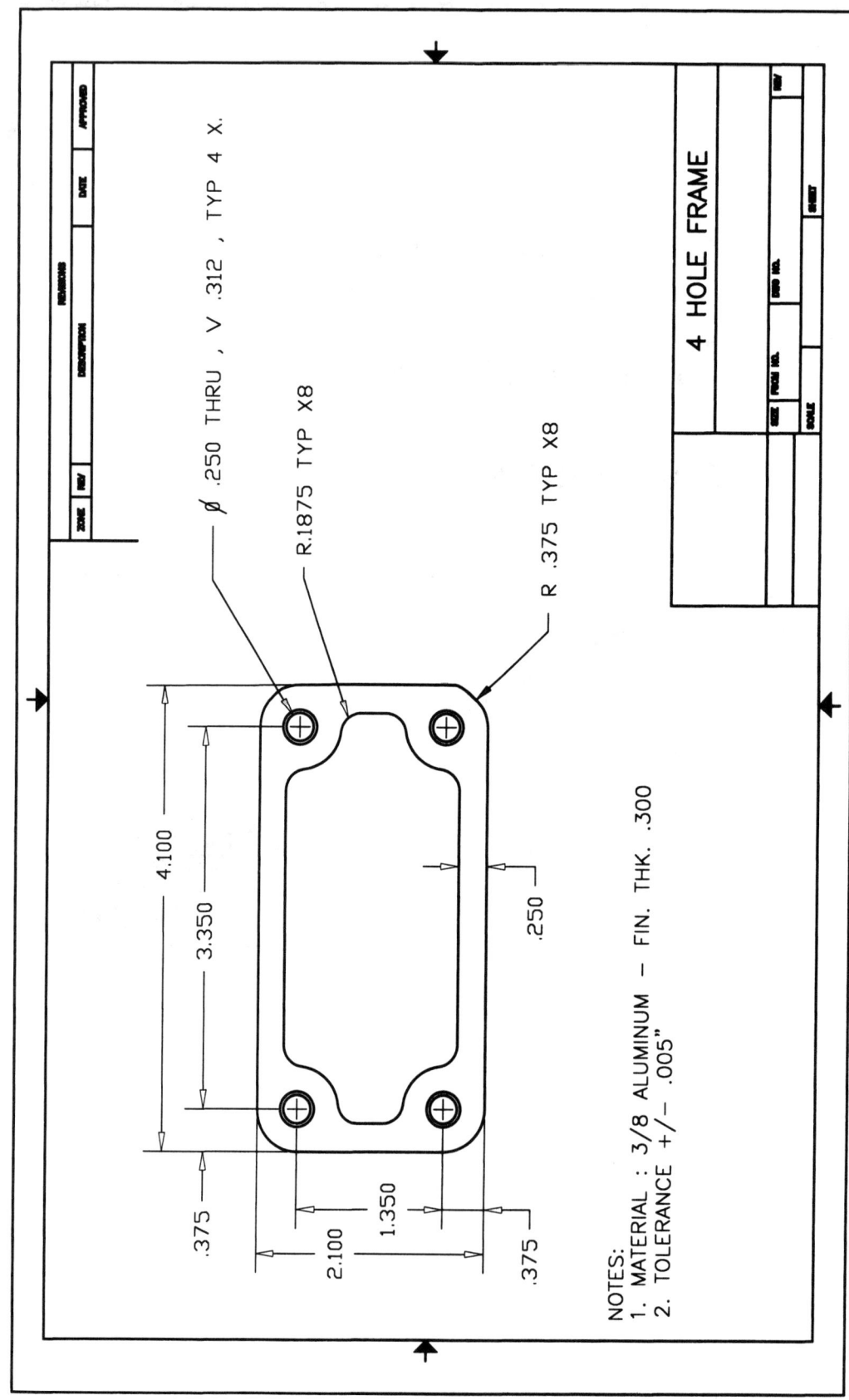

The blueprint for the Four-Hole Frame Tutorial

Chapter 6   SmartCAM Tutorial 5

Upon completion of this chapter, you should be able to:

- Define and use multiple workplanes in a process model.
- Succcessfully build a process model which will be machined on all surfaces.
- Construct, rough machine, and finish machine an irregular interior profile.
- Insert user events into the process model.
- Understand the correlation between the process model and the Display Modes dialog box.

# FOUR-HOLE FRAME

The purpose of this SmartCAM tutorial is to introduce the concepts of workplanes and user events. This project will more closely simulate "real world" applications in the respect that it will take you through the machining processes of all surfaces of a workpiece. In addition, you will incorporate additional workplanes so you will understand how to utilize SmartCAM in the event that you have workpieces stationed in multiple fixtures.

As with all process models, you must first have an active job operations file in order to successfully work within the SmartCAM environment. This can be a job file from a previous process model or it can be one that is built specifically for this model.

If you desire to use a job file from a previous process model, follow these steps:

1. Choose **File** from the menu bar.
2. Choose **Load Job File** from the menu selection.
3. Inside the **Load Job File** dialog box, choose the **File Select** button.
4. Select the appropriate directories and files to meet your needs.

If necessary, modify the existing job file to reflect the tools and steps in Table 6.1. If you choose to build a new job file specifically for this process model, use Table 6.1 to assist you.

## Constructing the Finish Profiles

When your Job Operations File is complete, you can begin constructing the finish profile. To do so, follow these steps:

**TABLE 6.1**

| Step # | Tool # | Type | Diameter | Speeds | Feeds |
|---|---|---|---|---|---|
| 1 | 1 | Face Mill | 3″ | 600 SFPM | .010 IPT |
| 2 | 2 | End Mill (2 flute roughing) | 5/8″ | 600 SFPM | .006 IPT |
| 3 | 3 | End Mill (4 flute finishing) | 5/8″ | 650 SFPM | .005 IPT |
| 4 | 4 | End Mill (2 flute roughing) | 5/16″ | 600 SFPM | .004 IPT |
| 5 | 5 | End Mill (4 flute finishing) | 5/16″ | 650 SFPM | .003 IPT |
| 6 | 6 | Spot Drill | 3/8″ (90 deg) | 600 SFPM | .006 IPR |
| 7 | 7 | Drill | 1/4″ | 600 SFPM | .004 IPR |

1. Choose **Insert** from the workbench.
2. Select the **After**, **Element Sequence**, and **With Step** options from the tool list.
3. Set the input fields of your **Insert** control panel to match those of Figure 6.1.
4. Choose **Geometry** from the workbench.
5. Choose the **Line** tool from the tool list.
6. Construct a line beginning at X 0.0, Y 0.0 and ending at X 0.0, Y 2.1.
7. Continue creating line geometry until you have a 2.1″ by 4.1″ rectangle as shown in Figure 6.2.

Make this rectangle the active group.

1. Choose **Group** from the workbench.
2. Choose the **Select All** button.

Edit the existing profile to construct the 3/8″ radius arcs.

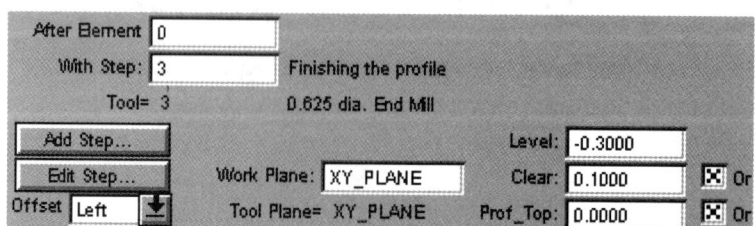

**FIGURE 6.1**
The input fields of the **Insert** control panel

**FIGURE 6.2**
The completed rectangle

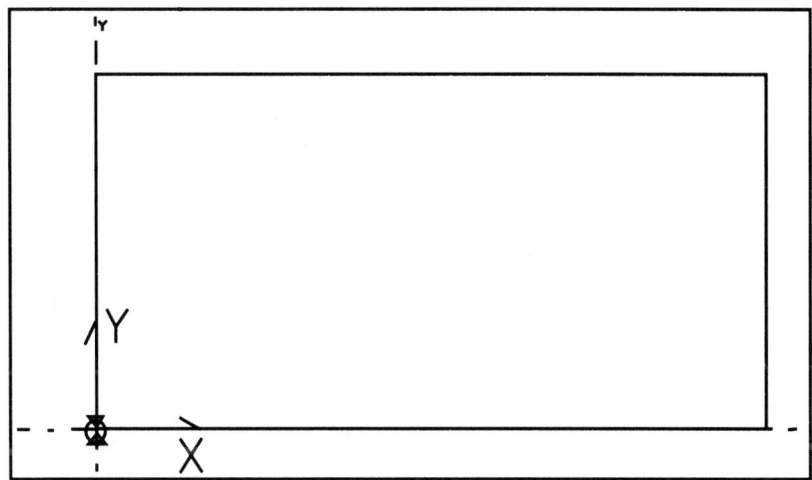

1. Choose **GeoEdit** from the workbench. (If **GeoEdit** is not on the workbench, it can be found under the **Edit** menu.)
2. Once the **GeoEdit** toolbox is open, select the **Blend** option.
3. Input .375" into the **Grp Outside Radius** input field.
4. With your mouse, choose the **Group Blend** button. All four corners of the active group will be blended simultaneously as reflected in Figure 6.3.

The next task is to put a lead in/lead out move at the midpoint of the left vertical line (element #1).

1. First split the element by choosing **GeoEdit** from the workbench.
2. Choose the **Split** tool from the tool list.
3. Choose **Element Division** as the method by which you split the element.

**FIGURE 6.3**
The 3/8" blended radii

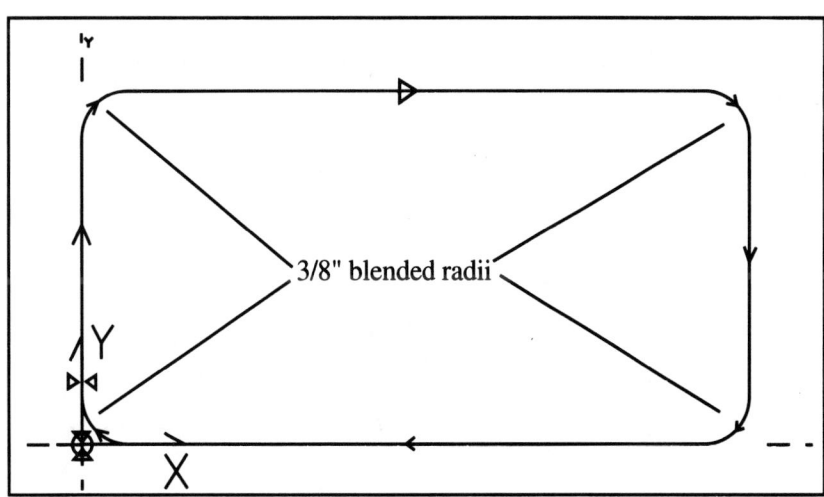

4. Accept the default of .5 for the **% Length**.
5. Highlight the **Select Split Element** input field by choosing it with your mouse.
6. Pick element 1 with your mouse. Element 1 should be the left vertical line.

**Remember:** *In situations where you are required to choose a particular element you can input that element in one of three ways. After the appropriate field is highlighted, you may choose the element from the graphics working area with your mouse, you may type the element number in with the keyboard, or you may select the element from the database list. Each method will work equally well.*

7. Choose **Lead In/Lead Out** from the tool list.
8. Select **Both** to create both a lead in and a lead out move.
9. Choose the **Arc** input field and input an **Angle** of 90.0 degrees and a **Radius** of 3/4".
10. Turn on the **Change Start** option.
11. Highlight **Select Element in Profile**.
12. Choose the upper portion (above the midpoint) of the left vertical line.

The lead in/lead out geometry will be inserted into your process model as shown in Figure 6.4.

Continue the process model by constructing the interior, finish profile, following these steps:

1. Choose **Insert** from the workbench.
2. Select the **After, Step Sequence,** and **With Step** options from the tool list.
3. Match the input fields of your **Insert** control panel with those of Figure 6.5.

Since the part has a 1/4" frame, the easiest method to create the interior profile is to simply create a wall offset of the existing geometry.

**FIGURE 6.4**
The lead in/lead out arcs

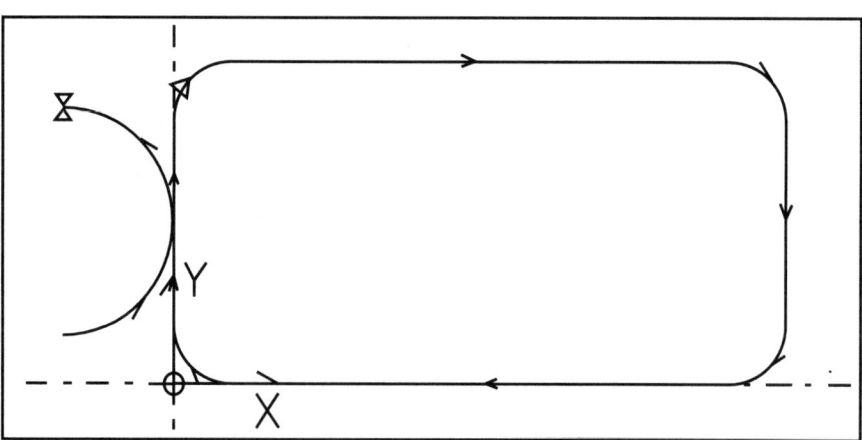

**FIGURE 6.5**
The input fields of the Insert control panel

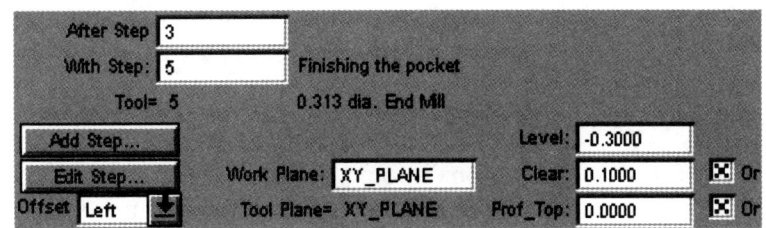

The exterior profile should still be grouped. If it is not, group the exterior profile before proceeding. Next, follow these steps:

1. Choose **Group** from the workbench.
2. Turn the **Remove** option on.
3. Choose **Element** from the tool list.
4. With your mouse, select the lead in/lead out arcs from the graphics work area.

    You do not want the lead in/lead out arcs to be offset.
5. Choose **Geometry** from the workbench.
6. Select the **Wall Offset** tool from the tool list.

Inside the **Wall Offset** control panel:

1. Set the **Wall Side** selector to **Right**.
2. **Distance** should be set to .250".
3. With your mouse, select the **Group Wall** button.

At this time, a rectangular profile should be on the inside of the original exterior profile, and your process model should match Figure 6.6.

Before proceeding with the process model, analyze the interior profile to verify the accuracy of the properties of each element.

**FIGURE 6.6**
The interior profile

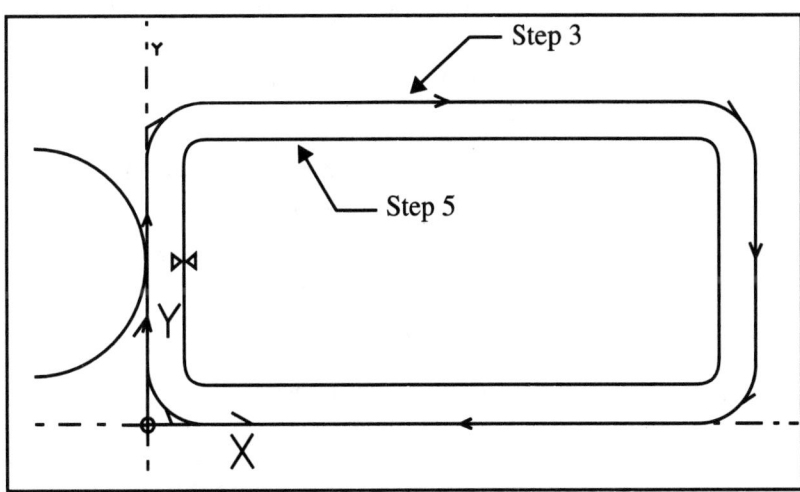

1. Choose **Utility** from the menu bar.
2. Choose the **Element Data** option.
3. With your mouse, systematically choose each element of the interior profile from the graphics work area. The **Element Data** dialog box should reveal a Z level of -.300", a profile top of 0.00, and a clear of .100" for each element. If any corrections need to be made, follow these steps:
   a. Remove all elements from the active group.
   b. Place the elements which need to be corrected in the active group.
   c. Choose **Edit** from the menu bar.
   d. Choose the **Property Chg** option from the list.
   e. Choose the **Toolpath** option.
   f. Upon opening the **Toolpath Property Change** dialog box, change only those values that are incorrect.
   g. Select **Accept** to make the changes.

Additionally, the direction of progression of the elements and the tool offset will be incorrect. They will be the same as those of the exterior profile. To correct these problems, follow these steps:

1. Choose **Group** from the workbench.
2. With your mouse select the **New Group** button.
3. Select the **Profile** option from the tool list.
4. With your mouse select the interior profile from the graphics work area.
5. Choose **Edit** from the menu bar.
6. Select the **Order Path** toolbox.
7. Choose the **Reverse Order** option from the tool list.
8. Inside the **Reverse Order** control panel, turn the **Direction Only** switch on.
9. Choose the **Group Reverse** button.

To correct the tool offset, follow these steps:

1. Again, choose **Edit** from the menu bar.
2. Again, select the **Property Change** option.
3. Select **Toolpath**.
4. Inside the **Toolpath Property Change** control panel, toggle through the **Offset** selections until the **Left** offset appears.
5. Choose **Accept** to complete the changes.

Before proceeding, verify the corrections:

1. Choose **Utility** from the menu bar.
2. **Element Data** should reveal a left offset, a Z value of -.300, a profile top value of 0.00, and a clear of .100.

The grouping arrows should also reveal a counterclockwise progression of the internal geometry.

Before you place the .375" radius in the corners of the internal profile, you must first delete the small radius blends that are currently in the corners. These radii are unnecessary and are simply a by-product of the wall offset operation that was done previously.

To delete these small radii:

1. Choose **GeoEdit** from the workbench. If it is not on the workbench, it can be found under the **Edit** menu selection.
2. Choose the **Delete** tool from the tool list.
3. With your mouse, pick each of the four small internal radii from the graphics work area.

To place the .375" radii in the internal corners:

1. Choose **Insert** from the workbench.
2. Choose the **After** and the **Element Sequence** tools from the tool list.
3. Turn on the **Match Element** option by selecting it with your mouse.
4. With your mouse, select the top line of the internal profile from the graphics work area. This will set all input fields of the current insert control panel to match those of the existing pocket finish profile.
5. Choose **Geometry** from the workbench. If **Geometry** is not on the workbench it can be found under the **Create** menu of the menu bar.
6. Choose the **Arc** tool of the tool list.
7. Inside the **Arc** control panel, set **Arc Direction** to **CW** (clockwise).
8. Input .375" for the radius.
9. Choose **Tangent El S:** and choose the top horizontal element of the external profile.
10. Next, choose **Tangent El E:** and choose the left vertical element of the external profile. Remember to choose the start and end points in a clockwise manner.

Your process model should now match Figure 6.7.

Place the arc in the lower left corner of the model:

1. Choose **Insert** from the workbench.
2. **After, Element Sequence, Match Element**, and **With Step** should all default from the previous set of steps.
3. Additionally, inside the **Insert** control panel, the **After Element** input field should be highlighted.
4. With your mouse, select the left vertical line of the interior profile according to Figure 6.8.

**FIGURE 6.7**
The first 3/8" corner radius

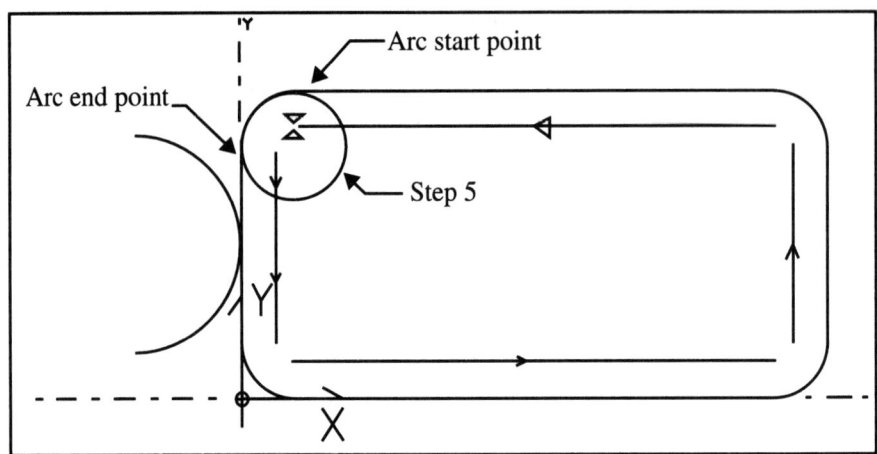

**FIGURE 6.8**
The pickpoint for the second arc

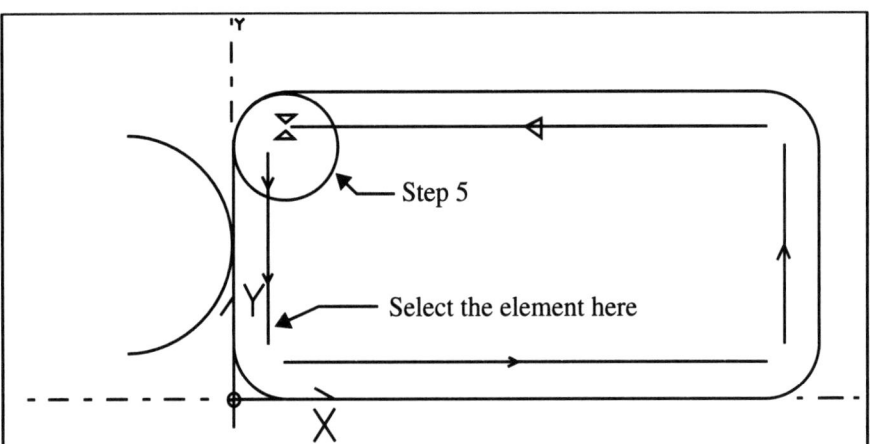

5. Choose **Geometry** from the workbench.
6. Choose the **Arc** tool of the tool list.
7. Inside the **Arc** control panel, set **Arc Direction** to **CW** (clockwise).
8. Input .375" for the radius.
9. Choose **Tangent El S:** and choose the left vertical element of the external profile.
10. Next, choose **Tangent El E:** and choose the bottom horizontal element of the external profile. Remember to choose the start and end points in a clockwise manner.

Your process model should now match Figure 6.9.

Continue placing the .375" radius arcs in the corners until all four arcs are in place. Once all four arcs are in place trim all excess geometry:

1. Choose **GeoEdit** from the workbench.
2. Choose the **Trim/Extend** tool from the tool list.

**FIGURE 6.9**
The correct location of the second arc

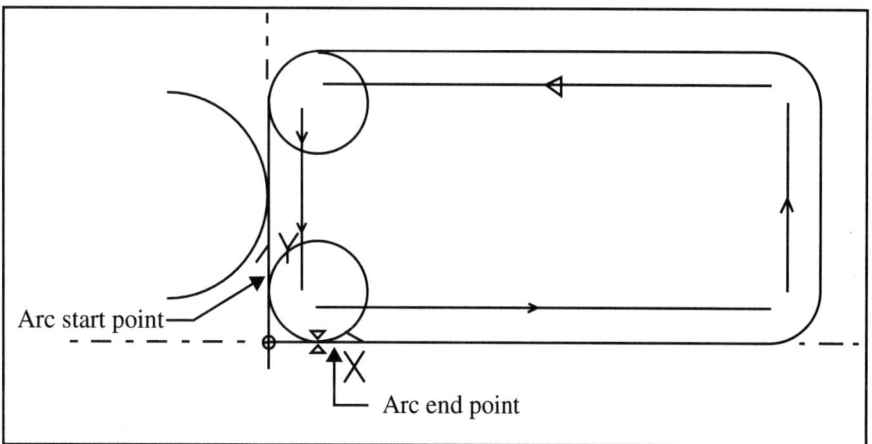

**Remember:** *With SmartCAM's **Trim/Extend** function, you must pick the portion of the element you wish to keep.*

After all the .375″ arcs are located properly and the elements are properly trimmed, your process model should match Figure 6.10.

Continue building your process model by placing the .1875″ radii in the internal corners.

1. Choose **GeoEdit** from the workbench.
2. Select the **Blend** tool from the tool list.
3. Input .1875″ in the **Inside Radius** input field.
4. With your mouse, highlight **Select 1st Element** by selecting it.
5. Proceed around the internal geometry, choosing the element blend points until all intersections are properly blended. Figure 6.11 provides further explanation regarding the mouse pickpoints.

Upon completion of the process of inserting all .1875″ radii, your process model should match Figure 6.12.

**FIGURE 6.10**
The completed insertion of the .375″ arcs

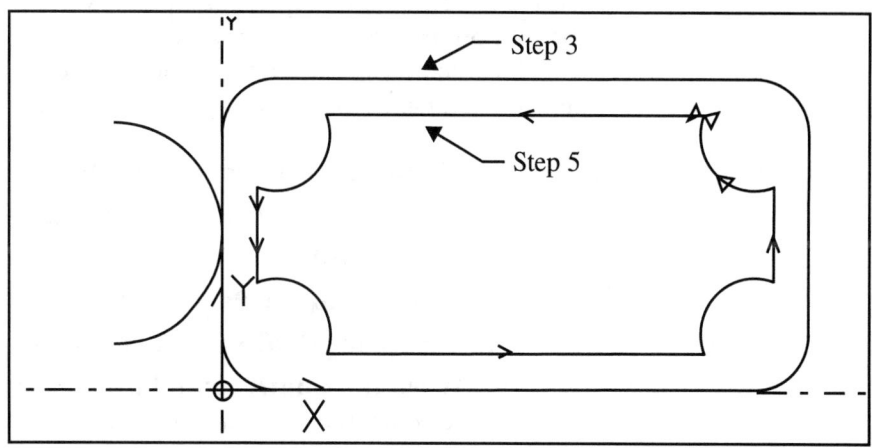

**FIGURE 6.11**
The mouse pickpoints

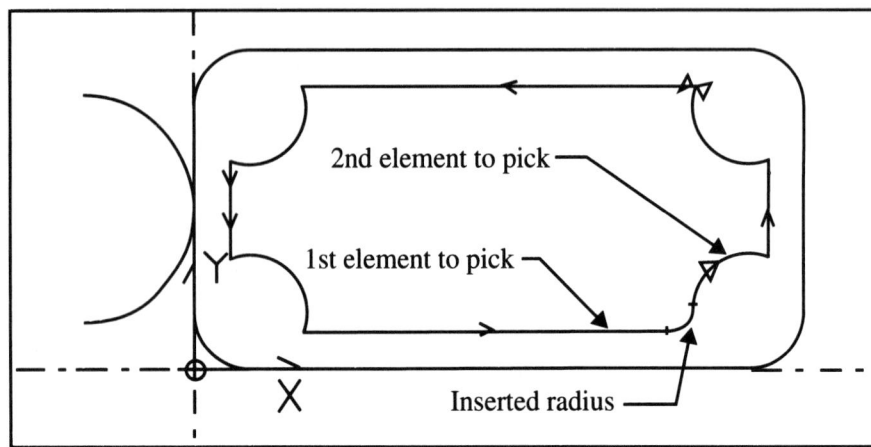

**FIGURE 6.12**
The completed pocket profile

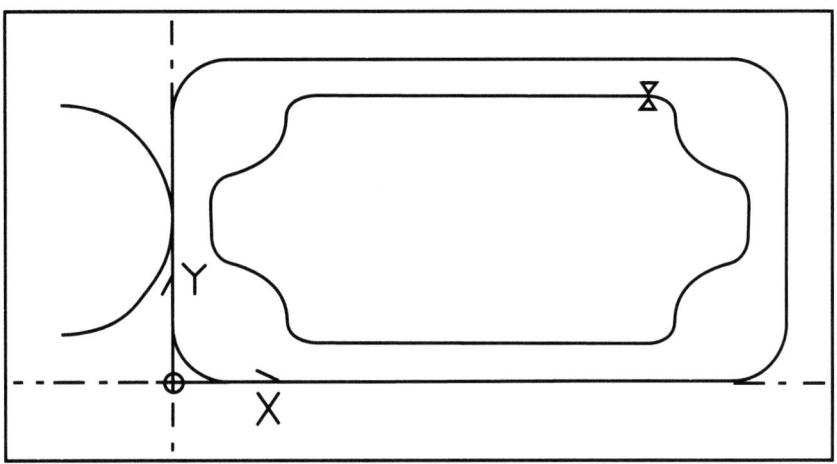

At this point the interior profile is complete. There may, however, be a few minor problems with the sequence of the profile. Run the **Show Path** function of SmartCAM to inspect the sequence of the geometry. When **Show Path** is running you have to look closely, so slow the speed down to about two. Remember that the solid lines of **Show Path** represent feed moves, while the dashed lines represent rapid traverse moves. The objective is to have the tool progress around each profile without interruptions.

If by chance the interior profile is out of sequence, the problem needs to be corrected before proceeding. Do this by following these steps:

1. Choose **Group** from the workbench.
2. Select the **Step** tool and choose step 5.
3. Choose **Edit** from the menu bar.
4. Select the **Order Path** toolbox.
5. Choose the **Chain** tool from the tool list.
6. When the **Chain** control panel opens, make sure that the **Chain** selector switch is turned on (the circle is filled-in with a black dot).

7. With your mouse, highlight the **Select an Element in profile** input field.
8. Pick the bottom horizontal line of the interior profile.

This series of steps will accomplish two objectives for your interior profile. First, it places all elements of the profile in sequential order. Second, it resequences the database so the element of the profile that was selected becomes the first element of the profile. This will help you when you insert your lead in/lead out geometry.

Again, analyze the database list and re-run **Show Path**. If the problem still exists, look at your geometry in all views to determine if all geometry is at the same Z level. The chain function will only work with geometry at the same Z level. The **Element Data** function of SmartCAM is also very useful in determining the Z level of geometry.

**Caution:** *It is very important to make sure all geometry is correct before proceeding to the next step.*

## Constructing the Roughing Profiles

After you have verified that both interior and exterior finish profiles are correct, the external roughing profiles can be created.

To create the external roughing profiles, follow these steps:

1. Choose **Group** from the workbench.
2. Select the **New Group** button to degroup the active groups.
3. Choose **Step** and select step 3.
4. Choose **Insert** from the workbench.
5. Select the **Before, Step Sequence,** and the **With Step** options from the tool list. **Match Element** should be off.
6. Fill in the input fields of your **Insert** control panel to match those of Figure 6.13.
7. Choose **Geometry** from the workbench.
8. Select the **Wall Offset** tool from the tool list.
9. Set **Wall Side** to Left.
10. Input .010" into the **Distance** input field.
11. Choose the **Group Wall** button.

**FIGURE 6.13**
The input fields of the **Insert** control panel

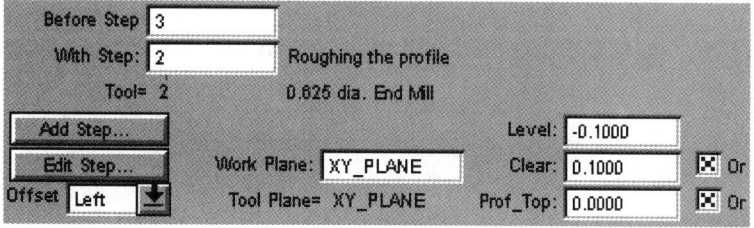

**FIGURE 6.14**
The outcome of the wall offset command

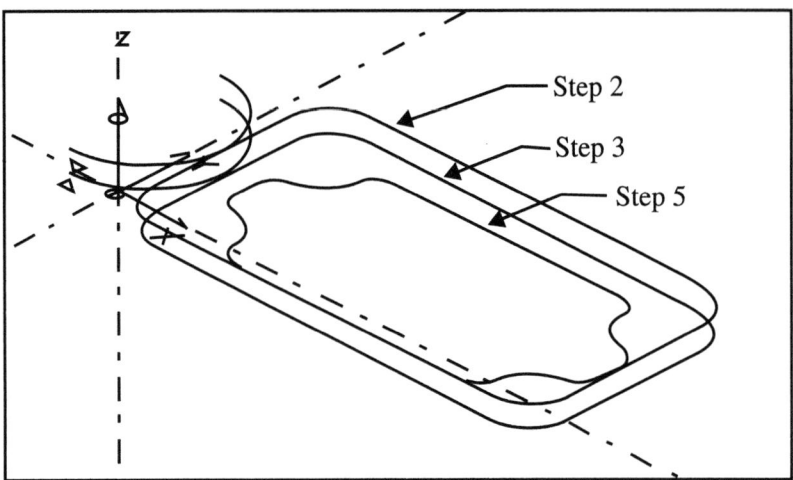

Upon completing step 11, your process model should match Figure 6.14.
To complete the external roughing process, group step 2, follow these steps:

1. Choose **Group** from the workbench.
2. Select the **New Group** button.
3. Choose the **Step** tool from the tool list.
4. Choose step 2.
5. Choose **Edit** from the menu bar.
6. Select the **Transform** toolbox.
7. Choose the **Move** tool.
8. Inside the **Move** control panel, turn **Copy** on.
9. Set **Copies** to 2.
10. With your mouse, select the **From 0** button.
11. For the **To Point** input field, enter X 0.0, Y0.0, and Z -.100.

Upon entering the Z value, the other two roughing profiles will be created. Your process model should now match Figure 6.15.

This will complete the external roughing and finishing operations. Continue the construction of your process model by roughing the internal pocket.

1. Obtain a full scale, top view of your process model.
2. Choose **Insert** from the workbench.
3. Choose the **Before**, **Step Sequence**, and **With Step** options from the tool list.

Fill in the input fields of your **Insert** control panel to match those of Figure 6.16.

1. Choose **Process** from the menu bar.
2. Select the **Rough** toolbox.

Chapter 6 SmartCAM Tutorial 5

**FIGURE 6.15**
The completion of the exterior roughing profiles

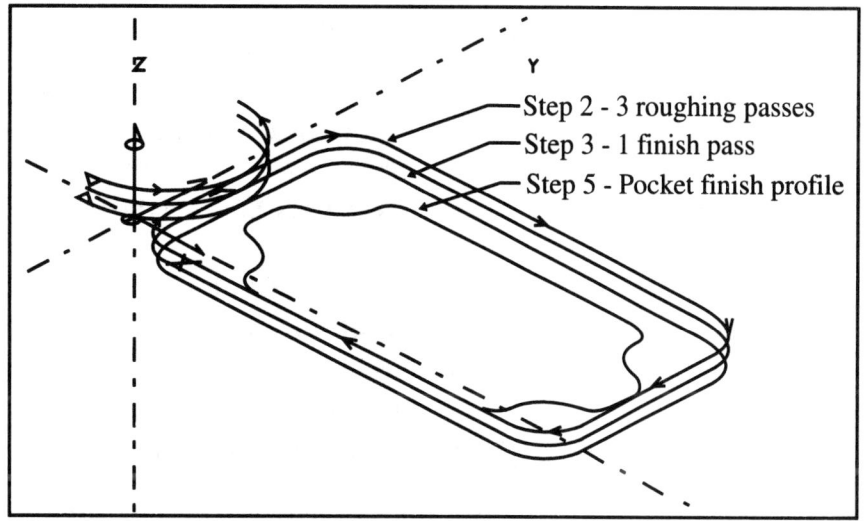

**FIGURE 6.16**
The input fields of the **Insert** control panel

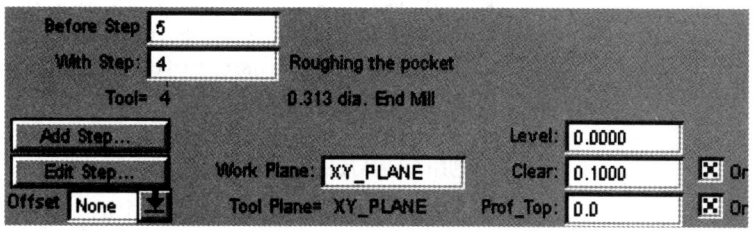

**FIGURE 6.17**
The input fields of the **Pocket** control panel

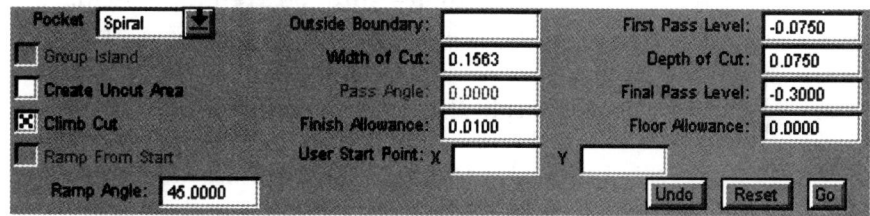

3. Choose the **Pocket** tool.
4. Fill in the input fields of your **Pocket** control panel to match those of Figure 6.17.

   For the **Pocket** input field the default setting **Spiral** was accepted. Both Linear and Zig Zag will work equally well for this example.

   **Outside Boundary** was intentionally left blank for this example. It may vary from process model to process model due to the sequence of the database list. It, however, will always be the finish profile of the pocket. (Remember, the Z level of the finish profile must be -.300″ and the profile top must be set to 0.0 for the roughing process to work correctly.)

   The **Width of Cut** default value of .1563″ (half the endmill diameter) was accepted for this input field. Other values will work. A smaller value will be easier on your endmill, but the process will take longer. Larger values will speed up your production rate, but it is harder on your tools.

The **Finish Allowance** is an appropriate amount that you wish to leave on the wall of the pocket for finishing purposes.

**First Pass Level** is the negative depth of the first roughing pass.

**Depth of Cut** is the positive depth of all successive passes.

**Final Pass Level** is the negative value of the final depth of the pocket.

5. Accept the default value for all other input fields.
6. With your mouse, select the **Go** button.

At this time, the pocket roughing routine is complete. Your process model should now match Figure 6.18.

With the completion of the roughing routine, you are now ready to complete the finish profile by inserting the **Lead In/Out** moves. It will be easier to work with the finish profile if you first mask the pocket roughing routine.

1. Choose **Utility** from the menu bar.
2. Select the **Show/Mask** option
3. Inside the **Show/Mask** control panel, select the **Hide** option.
4. Input 4 into the **Step** input field.
5. Obtain a full-scale, top view of the process model.

Next, split the lower horizontal line of the pocket finish profile:

1. Choose **Edit** from the menu bar.
2. Select the **GeoEdit** toolbox.
3. Choose the **Split** tool from the tool list.
4. Choose the **Element Division** option.

**FIGURE 6.18**
The completion of the pocket roughing routine

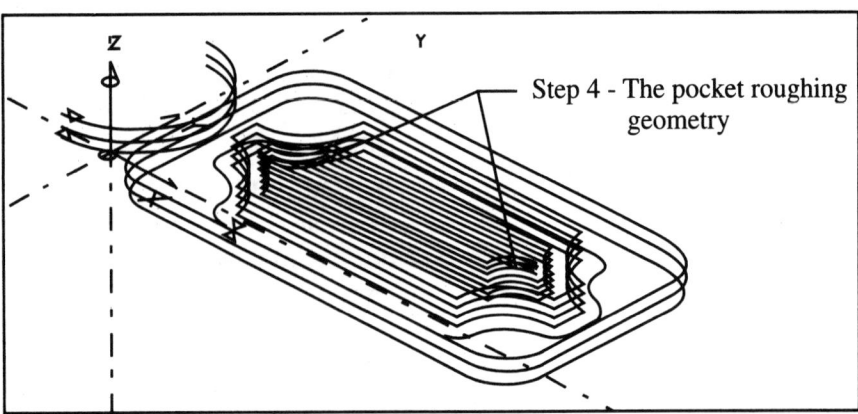

5. Accept the default value of .5 percent.
6. Choose **Select Split Element** and pick the lower horizontal line of the interior finish profile.
7. Choose **Lead In/Out** from the tool list.
8. Inside the **Lead In/Out** control panel, choose the **Both** option.
9. Select **Arc** as the type of lead in/lead out move.
10. Set **Angle to** 90.0.
11. Set **Radius** to .500".
12. Turn **Change Start** on.
13. Highlight **Select Element in Profile** and pick the right side of the lower horizontal line. (Pick the element between the midpoint of the line and the .1875" radius.)

With the pocket roughing elements unmasked, your process model should now match Figure 6.19.

## Spot Drilling and Drilling the Holes

The next step in building your process model is the construction of the four spot drilled and drilled holes.

1. Choose **Insert** from the workbench.
2. Select the **After**, **Step Sequence**, and **With Step** tools from the tool list.
3. Fill in the input fields of your **Insert** control panel to match those of Figure 6.20.
4. Choose **Geometry** from the workbench.
5. Choose the **Hole** tool from the tool list.
6. Fill in the input fields of your **Hole** control panel to match those of Figure 6.21.

**FIGURE 6.19**
The lead in/lead out arcs of the interior finish profile

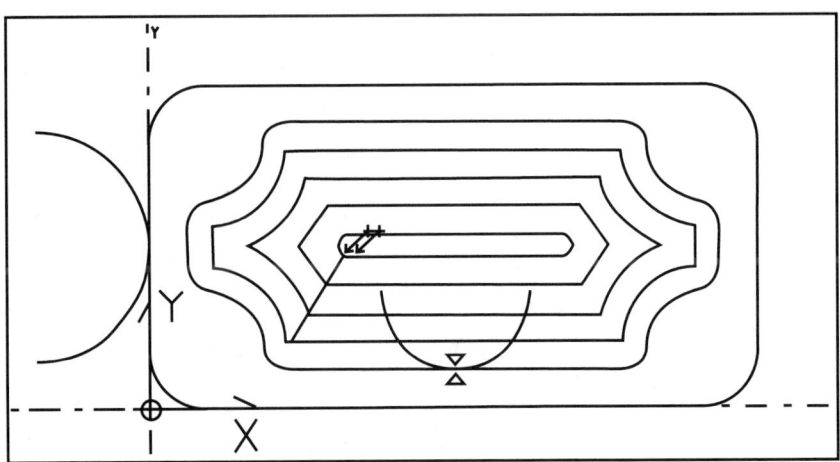

**FIGURE 6.20**
The input fields of the **Insert** control panel

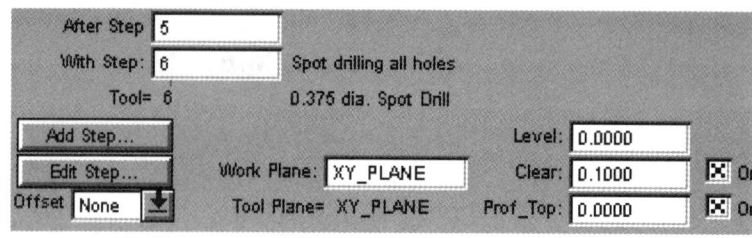

**FIGURE 6.21**
The input fields of the **Hole** control panel

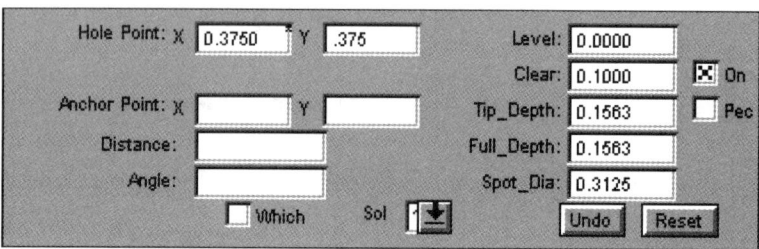

**Spot Dia** is the value taken from the print. Input this value first and the values of **Tip Depth** and **Full Depth** are automatically calculated from the job file information.

**Hole Point** values are the hole locations taken from the print.

Next, you will want to place the drilled hole on the model.

1. Choose **Insert** from the workbench.
2. Select the **After**, **Step Sequence**, and **With Step** tools from the tool list.
3. Set **After Step** to 6.
4. Set **With Step** to 7.
5. **Offset** should be set to None.
6. **Level** should be set to 0.000.
7. **Clear** should be set to .100".
8. **Profile Top** should be set to 0.000.
9. Choose **Geometry** from the workbench.
10. Select the **Hole** tool from the tool list.
11. Set **Full Depth** to .300".
12. Highlight **Hole Point** and pick the spot drill hole from the graphics working area. Verify that the Snap Point icons and the Edit filters are set to allow the picking of hole elements.

Your database list should now reveal two hole elements. Additionally, the isometric view should show geometry representing a spot drill and a drill.

The next step is to transform the spot drill and the drill to the other locations:

1. Choose **Group** from the workbench.
2. With your mouse, select the **New Group** button.

3. Choose the **Step** tool.
4. Select the Spot Drill and the Drill (steps 6 and 7).
5. Choose **Edit** from the menu bar.
6. Choose the **Transform** toolbox.
7. Select the **Move** tool from the tool list.
8. **Copy** should be turned on.
9. **Copies** should be set to 1.
10. **Sort by Tools** should be turned on.
11. With your mouse, select the **From 0** button.
12. For **To Point** input X 4.1 -(.375*2), Y0.00, Z0.00. Remember, since the input fields of SmartCAM will accept mathematical formulas, it is always better to enter the values straight from the print. The "X" value is simply the part length minus twice the radius.

Your process model should now match Figure 6.22.
To create the two remaining holes:

1. Choose **Group** from the workbench.
2. Again, select the **Step** tool and pick steps 6 and 7.
3. Choose **Transform** from the workbench.
4. Select the **Move** tool from the tool list.
5. **Copy** should be turned on.
6. **Copies** should be set to 1.
7. **Sort by Tools** should be turned on.
8. With your mouse, select the **From 0** button.
9. For **To Point** input X 0.0, Y 2.1-(.375*2), Z0.0

At this point, the four spot drilled and the four drilled holes should be complete. The database list and the geometry in the graphics work area should reflect the holes.

**FIGURE 6.22**
All irrelevant geometry has been masked

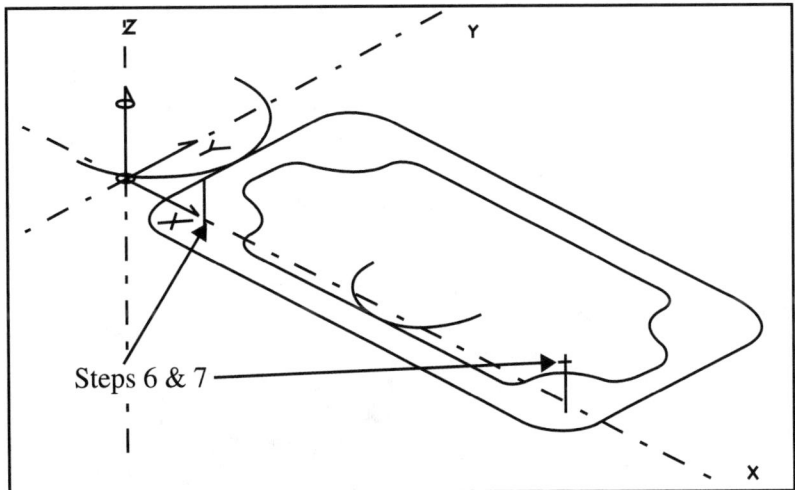

# Facing the Workpiece

At this point of the process model, you are ready to construct the facing cycle. This facing cycle will be done a little differently from the previous examples. Assume that your blank stock starts out at .375" and you need to face .075" to bring the part to size. You will insert three facing cuts, one on the first side, and two on the second side. An "M00" program stop will be inserted between the two sides.

To do so, follow these steps:

1. Choose **Insert** from the workbench.
2. Select the **Before**, **Step**, and **With Step** tools from the tool list.
3. Fill in the input fields of your **Insert** control panel to match those of Figure 6.23.
4. Choose **Geometry** from the workbench. If **Geometry** is not on the workbench, it can be found under the **Create** menu.
5. Select the **Line** tool from the tool list.
6. Create a line with a **Start Point** of X-2.00, Y1.05, and Z -.025.
7. Input an **End Point** of X6.1, Y 1.05, and Z-.025.

This will create one line of geometry that is assigned to the facing step.

8. Choose **Group** from the workbench.
9. Select the **New Group** button to degroup the active group.
10. Select the **Step** tool from the tool list.
11. Choose step 1.
12. Choose **Transform** from the workbench.
13. Select the **Move** tool from the tool list.
14. Turn **Copy** on.
15. Set **Copies** to 2.
16. With your mouse, select the **From 0** button.
17. Input X0.00, Y0.00, and Z-.025 for the **To Point** input field.

This will create the additional two facing passes.

At this time the construction of the facing passes is complete. However, the objective is to be able to face both sides of the workpiece. In order to achieve this objective, you must have the capabilities to stop the machining cycle and flip the part immediately after the first pass.

**FIGURE 6.23**
The input fields of the **Insert** control panel

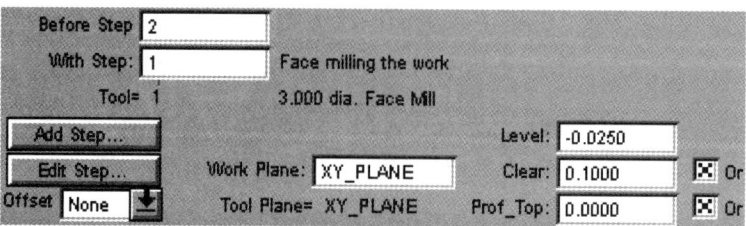

**FIGURE 6.24**
The input fields of the **Insert** control panel

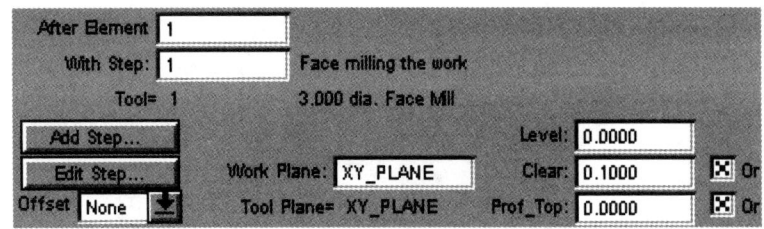

**FIGURE 6.25**
The input fields of the **User Event** control panel

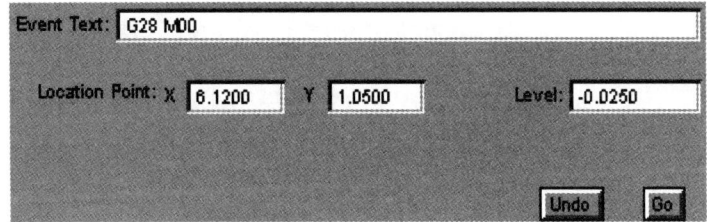

To accomplish this task, follow these steps:

1. Choose **Insert** from the workbench.
2. Select the **After**, **Element Sequence**, and **With Step** options from the tool list.
3. Fill in the input fields of your **Insert** control panel to match those of Figure 6.24:
4. Choose **Create** from the menu bar.
5. Choose the **User Elmts** toolbox from the list.
6. Select the **User Event** tool from the tool list.
7. Fill in the input fields of your **User Event** control panel to match those of Figure 6.25.

    The **Event Text** is the machine "G" code, "M" code, or note to the operator that you wish to input. You will need to be familiar with your machine codes to take full advantage of this toolbox.

    Neither the **Location Point** nor the **Level** are critical values. These are simply locations in the graphics work area where a "User Event" representation will be placed. Theoretically, any value could be placed here and the end result would be the same. For simplicity's sake, each of these input fields has been highlighted and the end point of the first facing pass has been picked with the mouse.

8. With your mouse, choose the **Go** button.

    The **User Event Text** will be input into the database list.

The "G28" code was input into the database so the tool would clear the workpiece to enable moving the part. You must now insert a **Point/Rapid** move so that the tool will rapid traverse back to the proper location. Failing to insert the **Point/Rapid** move will cause the tool to feed from the machine zero position rather than rapid traverse.

**FIGURE 6.26**
The input fields of the Point/Rapid control panel

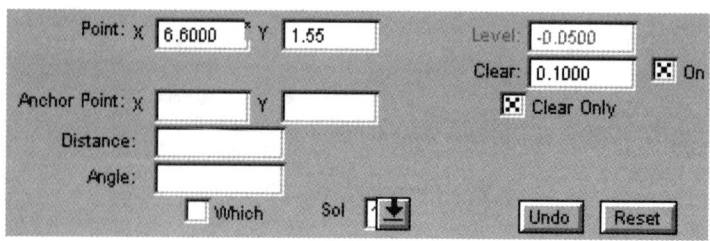

All **Insert** properties will remain the same as for the **User Event**.

1. Choose **Geometry** from the workbench.
2. Select the **Point/Rapid** tool from the tool list.
3. Fill in the input fields of your **Point/Rapid** control panel to match those of Figure 6.26.

    **Point** is the location you wish the tool to rapid traverse to.

    For this example, turn **Clear** on and also turn the **Clear Only** switch on.

Upon completion of the Point/Rapid control panel, a Point/Rapid move will be inserted into your database list.

There is a possibility you may have problems with the **Point/Rapid** function of SmartCAM, depending upon the setup of your machine define and template files. You will need to process code and then analyze the code to determine if you have a problem.

In the event that your code is incorrect, there are several things you can do to increase the possibility of correct code.

1. The Point/Rapid location **Point** must be X and Y values which are different from the start point of the next element. The end result of values which are the same is that the machine will, after the G28, rapid traverse to Z0.0 and then feed to the X,Y location of the second facemill pass.
2. It may be feasible to eliminate the G28 user element altogether and replace it with an arbitrary point/rapid location to move the tool away from the part. You will, however, need to keep the M00 command so that all motion is stopped. You will also need to insert a point/rapid command to move the tool back to the cutting position. Remember, the Point/Rapid location **Point** must be X and Y values that are different from the X and Y starting point of the next element.
3. Analyze your code. If the last "G" code before the G28 is a rapid move, the next X,Y location after the G28 will also be in rapid traverse. (G00 is modal and of a different family from G28). If, however, the last "G" code before the G28 is a G01 move, the next X,Y location after the G28 will also be in linear feed. You may, with your instructor's supervision, need to modify the template file, if the latter is the case. A "G00" code will need to be added to the @rap section of the template file.

It is important to understand that the actual .PM4 portion of SmartCAM does not recognize the "G28 M00" that was input by the user event. Only the machine define and template files work with the user events. The .PM4 portion of SmartCAM does not recognize that the M00 code shut everything off. If your code generator is not equipped to handle such events, you will have to make special provisions to ensure the completeness of your "G" code.

In this example, you need to turn the spindle and the coolant back on after the "G28 M00". To accomplish this task, simply insert two more user elements. One will insert "S 763 M03" and another will insert an "M08." Depending upon individual machine and controller types, other functions may need to be considered.

Another point that is worth mentioning is the direction of the facemill passes. All of the facing passes start from the left side of the part and feed toward the right. This requires unnecessary rapid traverse moves. To reduce the rapid moves and to increase production, change the direction of the second facing cut. This will allow less rapid moves and more time engaged in metal removal.

# Construction of Workplanes

To make this part as productive as possible it is necessary to machine the work in two separate work holding fixtures. There are two ways that you can approach multiple workpiece machining in SmartCAM. In the first method, you simply copy the geometry the required distance either in the "X" or the "Y" axis. There is, however, a drawback to this method of multiple workpiece construction. The distance that the workpiece is transformed in SmartCAM must correspond exactly to the distance from workpiece to workpiece at the machine. For example, if the distance from the first part to the second part on the CAM system is eight inches, it **must** be eight inches from the first part to the second part at the machine tool.

This type of exactness is not always practical in machining due to restrictions of machine tool work envelopes and milling machine vise sizes. There is, however, a better way to deal with multiple workpiece machining.

The second way to work with multiple workpieces in SmartCAM is to define multiple workplanes. Geometry is first constructed on the "XY" workplane as normal. A separate workplane—a "G55" workplane for example—is then defined at an arbitrary location. You then move and assign the required geometry to the newly defined workplane. Unlike the previous example of multiple workpiece machining, the distance between multiple workpieces is completely irrelevant when machining on separate workplanes. At the machine tool, the different work holding fixtures are defined as the different workplanes.

In regard to the Four Hole Frame, the G54 workplane is the original "XY" workplane. It is a pre-defined "system workplane." This simply means that it was not defined by the user. You will do the face milling, the spot drilling, and the drilling at the G54 workplane. You will then define a G55 workplane and do all of the internal and external profile milling here. The G54 workplane will be a milling machine vise that will hold the part to face mill and drill. The G55 workplane will be a simple plate fixture that will hold the part by the previously drilled bolt holes.

To accomplish the task of defining a "G55" workplane, follow these steps:

1. Choose **Workplane** from the menu bar.
2. Select the **Define Plane** option from the list.
3. Fill in the input fields of your **Define Plane** dialog box to match those of Figure 6.27.

    **3 Points** must be turned on. This option allows you to define an **Origin Point** of your new workplane, a direction in which the positive "X" axis points, and a third direction in which the positive "Y" axis points.

    **Match Plane** must be turned on. It is a necessity that your **workplane Name** match your **Tool Plane** name. (After you fully define your G55 workplane, you will need to insert a G55 user event. If you attempt to generate code and your **Tool Plane** name is different from your **workplane Name** you could get improper "G" code).

    **Origin Point** is the origin of the new workplane. If the **From World** option is off (as it currently is in Figure 6.27), the values for **Origin Point**, **Plus X Point**, and **Third Point** are in relation to the active workplane. If the **From World** option is on (an x in the box reflects the on condition), the values for **Origin Point**, **Plus X Point**, and **Third Point** are in relation to the system defined XY workplane.

    **Plus X Point** is a coordinate value that has an X value greater than the X value that was entered into the **Origin Point** input field. The Y and Z values will be the same as those entered in the **Origin Point**. This will allow the X axis icon to point in the positive direction.

    **Third Point** is a coordinate value that has the same X value as that of the **Origin Point** and a Y value greater than that of the **Origin Point**. The Z value will remain the same as the previous two. This will allow the Y axis icon to point in the positive direction.

4. With your mouse, select the **Accept** button.

Upon accepting these values, the G55 workplane is created. The G55 workplane also becomes the active workplane. Any geometry created will now be input on the G55 workplane.

**FIGURE 6.27**
The input fields of the **Define Plane** dialog box

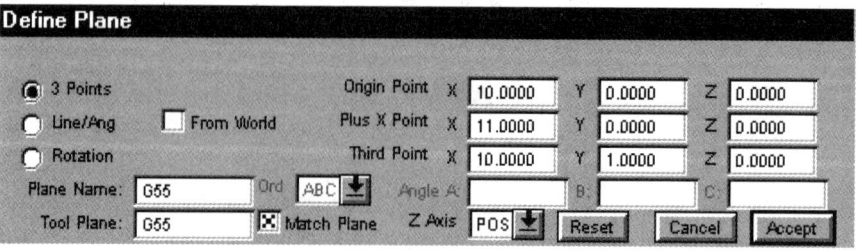

# Chapter 6 SmartCAM Tutorial 5

For further explanation of workplanes and active workplanes, follow these steps:

1. Choose **Insert** from the workbench.
2. Within the **Insert** control panel, pick the word **Workplane** with your mouse.

Figure 6.28 shows an example of a typical **Insert** control panel. For this example, the **Workplane** input field was chosen with the mouse. Upon the selection, the database list opened and revealed the currently defined workplanes. Figure 6.28 shows a dashed line beside the G55 workplane. The dashed line represents the "active" workplane or the workplane to which newly constructed geometry will be assigned.

Additionally, several workplanes show an "R" to the left of the workplane designation. The "R" indicates the reserved status. When a workplane is reserved, the physical properties of that workplane can neither be altered nor can the workplane be deleted.

As was explained earlier, once a workplane is defined it becomes the active workplane. In the graphics work area, active workplanes are shown by the workplane indicators.

Verify the "G55" workplane is the active workplane:

1. Choose **View** from the menu bar.
2. Choose the **Zoom** option.
3. Upon opening the **Zoom** dialog box, enter a **Zoom Magnification Factor** of .5.
4. When **View Center** highlights, select the extreme right portion of your process model from your graphics work area.

The workplane indicator, representing the G55 workplane, should now be seen as shown in Figure 6.29.

Once the "G55" workplane has been properly constructed, the next step is to move the appropriate geometry to the appropriate workplane.

To place the milling geometry on the G55 workplane, follow these steps:

1. Choose **Group** from the workbench.
2. Select the **New Group** button to deactivate the current group.
3. Choose the **Step** option.

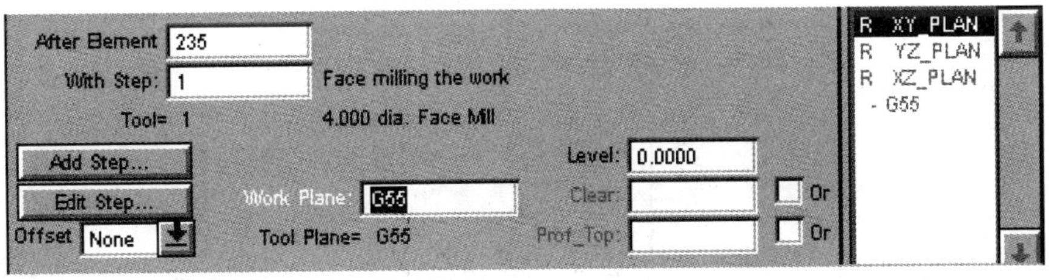

**FIGURE 6.28**
A typical Insert control panel

**FIGURE 6.29**
The workplane indicator representing the G55 workplane

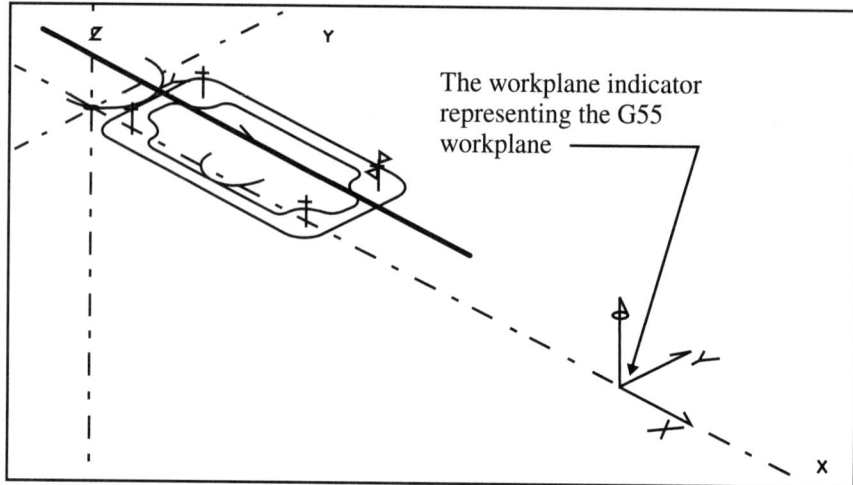

4. Choose steps 2,3,4, and 5.
5. Choose **Edit** from the menu bar.
6. Select the **Transform** toolbox.
7. Choose the **Move** tool.
8. Make sure **Copy** is turned off.
9. With your mouse, pick the **From 0** button.
10. Highlight the word "**Destination Plane.**"
11. Pick the G55 workplane from the database list.

The grouped elements will move to the G55 workplane. Your process model should match Figure 6.30.

To verify that the process model is correct, run Show Path and watch the simulation very carefully. It is quite possible that the steps are out of sequence. To correct this problem, follow these steps:

1. Choose **Group** from the workbench.
2. With your mouse, select the **New Group** button.

**FIGURE 6.30**
The geometry of the G54 and G55 workplanes

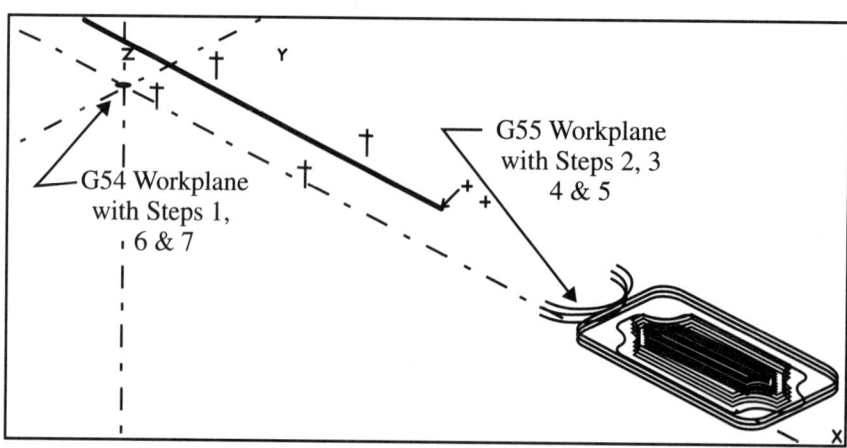

3. Select the **Step** option.
4. Choose Step 1, Step 6, Step 7, Step 2, Step 3, Step 4, and Step 5.
5. Inside the **Group** control panel, select the **Sequence Move** button. (Make sure the **By Selection Order** switch is turned on.)

This will properly sequence the steps in the order in which they should be machined. Again, run the Show Path function of SmartCAM to verify the corrections.

At this point your process model is complete except for two minor alterations. You must now insert a G55 code into your database so that all the elements on the G55 workplane will be coded out as G55 elements.

1. Choose **Insert** from the workbench.
2. Select the **Before**, **Step Sequence**, and **With Step** options from the tool list.
3. Input the following values into your **Insert** input fields:

   | | |
   |---|---|
   | **Before Step** | 2 |
   | **With Step** | 2 |
   | **Offset** | None |
   | **Work Plane** | G55 |
   | **Level** | 0.000 |
   | **Clear** | .100 |
   | **Prof Top** | 0.000 |

4. Choose **Create** from the menu bar.
5. Select the **User Elmts** toolbox.
6. Choose the **User Event** tool.
7. Input "G55" for the **Event Text**.
8. Choose the **Go** button.

The second minor alteration can be fixed by simply choosing **Utility > Display Modes**. Under the heading of "World Coordinates" turn **Element Display** off. This will affect the XYZ coordinates of the elements. With Element Display on, all elements will have XYZ coordinates based on their World XYZ location. This is incorrect for your applications. If you will turn this switch off, all XYZ coordinates will have values that are relative to the active workplane.

This concludes the tutorial on the Four Hole Frame. Many difficult concepts were covered in this chapter. It is imperative that you gain a complete understanding of the material covered in this chapter before you proceed. You may find it desirable to work through this tutorial more than once.

# CHAPTER 7

# SmartCAM Tutorial 6

## Support Bracket

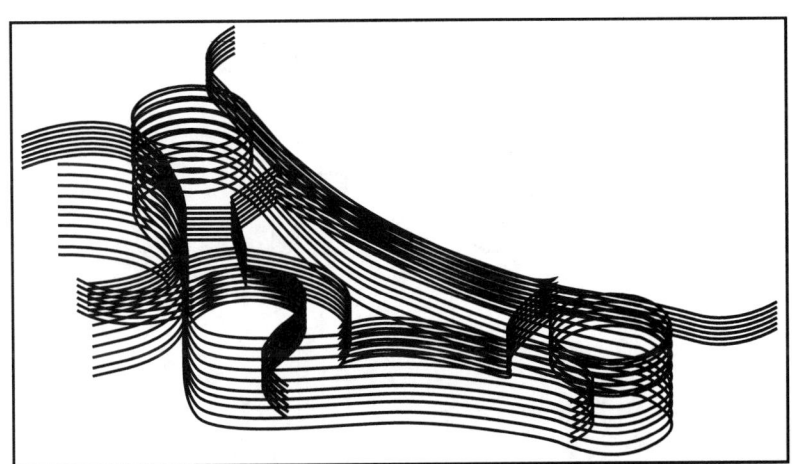

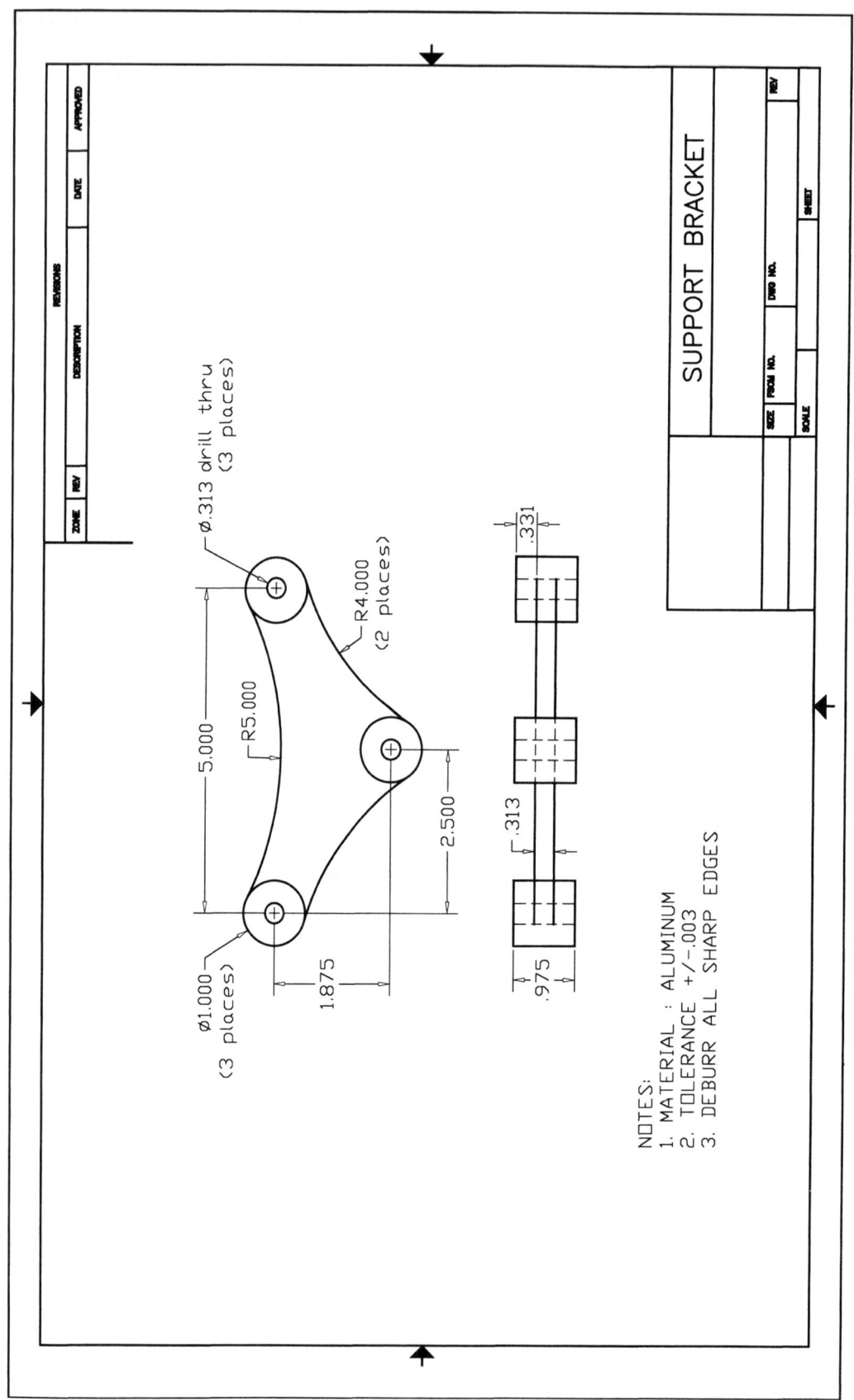

The blueprint for the Support Bracket Tutorial

Chapter 7   SmartCAM Tutorial 6

Upon completion of this chapter, you should be able to:

- Construct a job file "on the fly" or as the need arises for the tooling.
- Machine a workpiece on all sides which consists of open profiles and multiple islands.
- Construct a pocket roughing routine which will lead in to the profile as opposed to plunging into the profile.
- Manipulate the starting point of a pocket roughing routine.
- Identify and solve errors in a pocket roughing routine.

# SUPPORT BRACKET

In the previous tutorial, you were introduced to the concept of machining a workpiece on all surfaces. In order to accomplish this task, it was necessary to machine the workpiece in two different fixtures. It was also necessary to define a different workplane for each of those fixtures. Furthermore, in the first fixture, it was necessary to flip the work in order to face mill both sides.

In this tutorial you will continue with the processes necessary to machine your workpiece on all sides. You will also define multiple workplanes and shift the work to these workplanes in order to do the required machining. You will, however, add one additional step to this tutorial. In order to machine both sides of this part, it will be necessary to flip the part twice, once in the first fixture and once in the second fixture.

As with all SmartCAM operations, a valid job operations file must exist before you can accomplish any machining operation. However, the job file for this process model will be created a little differently than in past tutorials. For the sake of demonstrating a different method to create a job file, this job file will be created as the need arises for the steps.

## Creating the Finish Profile

To begin this tutorial you will construct the finish profile.

1. Choose **Insert** from the workbench.
2. Select the **After**, **Element Sequence**, and **With Step** tools from the tool list.

Upon the selection of **With Step**, an error message as shown in Figure 7.1 will appear.

3. Simply choose OK and the job planner will open to reveal the tooling and step selections as normal.

   For **Op Category** choose **Milling Operation**.
   For **Op Type** choose **Finish Milling**.

**FIGURE 7.1**
The error message that appears upon selection of **With Step**

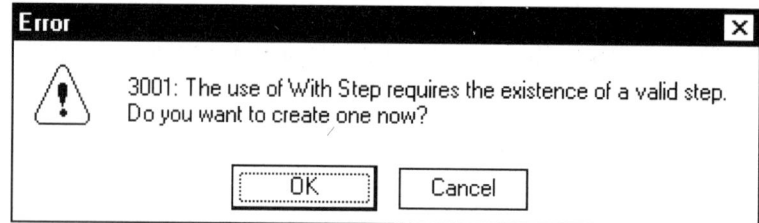

For **Tool Category** choose **Milling Tools**.
For **Tool Type** choose **Endmill**.

4. Continue setting up the finish step to reflect Table 7.1.

Once the process of entering step 3 into the job operations file is complete, continue with the **Insert** control panel.

Fill in the input fields of your **Insert** control panel to match those of Figure 7.2.

1. Choose the **Geometry** toolbox from the workbench.
2. Select the **Arc** tool from the tool list.
3. Fill in the input fields of your **Arc** control panel to match those of Figure 7.3

   A clockwise arc direction should be chosen to enable climb milling.

   The center point of X 0.0 and Y 0.0 is the location of the first arc.

   A radius of .500" should be chosen according to the blueprint.

4. Once these input fields are filled in, choose the **Full Arc** button.

   A complete circle should be shown in the graphics work area at the origin.

**TABLE 7.1**

| Step # | Tool # | Type | Diameter | Speeds | Feeds |
|---|---|---|---|---|---|
| 3 | 3 | End Mill (4 flute finish) | 3/4" | 600 SFPM | 30 IPM |

**FIGURE 7.2**
The input fields of the **Insert** control panel

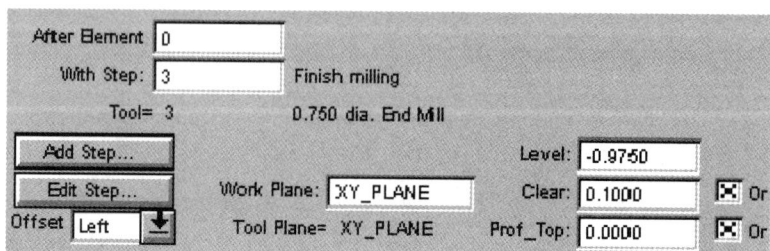

**FIGURE 7.3**
The input fields of the **Arc** control panel

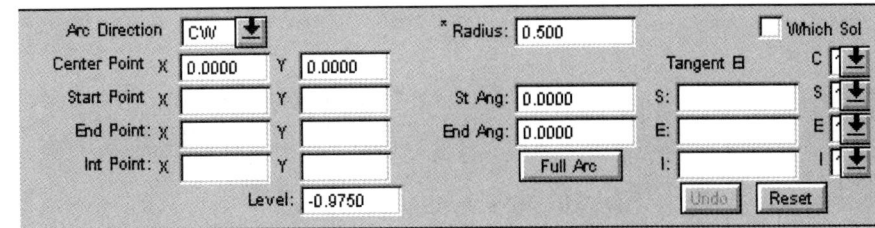

5. Choose **Group** from the workbench.
6. Select the **Element** tool and choose the circle.
7. Choose **Edit** from the menu bar.
8. Select the **Transform** toolbox.
9. Choose the **Move** tool from the tool list.
10. Turn **Copy** on and set **Copies** to 1.
11. With your mouse, choose the **From 0** button.
12. For **To Point**, input X 2.5, Y 1.875, Z 0.0.

Continue building your third arc:

1. Inside the **Move** control panel, again choose the **From 0** button.
2. For the **To Point** input field, input X -2.5, Y 1.875, and Z 0.0.

Three arcs should now be visible in the graphics work area as shown in Figure 7.4.

The next step in building your process model is constructing the three larger arcs.

1. Choose **Insert** from the workbench.
2. Select **After** and **Element Sequence** from the tool list.
3. Turn **Match Elmt** on. Recall from the previous tutorial that **Match Element** is a feature of SmartCAM that will automatically set the input fields of the **Insert** control panel to match those of the chosen element.

**FIGURE 7.4**
The completed 1″ diameter arcs

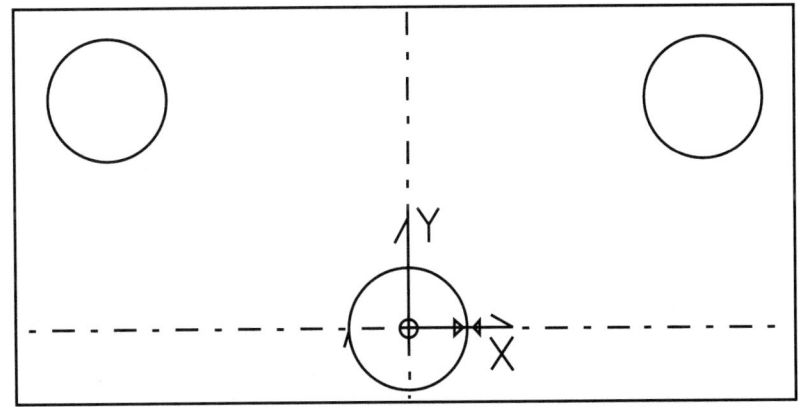

4. With your mouse, choose the upper left circle from the graphics work area.
5. Choose **Geometry** from the workbench.
6. Select the **Arc** tool from the tool list.

Inside the **Arc** control panel:

7. Set **Arc Direction** to **CCW**.
8. Set **Radius** to 5.00".
9. Choose the **Tangent El** start point (S:) as the upper right quadrant of the upper left circle.
10. Choose the **Tangent El** end point (E:) as the upper left quadrant of the upper right circle.

Your process model should match Figure 7.5. Additionally, Figure 7.5 further explains the mouse pickpoints described in steps 9 and 10.

Continue building the other two arcs:

1. Choose **Insert** from the workbench.
2. Select **After** and **Element Sequence** from the tool list.
3. **Match Element** should still be on.
4. With your mouse, choose the upper right arc from the graphics work area.
5. Choose **Geometry** from the workbench.
6. Select the **Arc** tool from the tool list.

Inside the **Arc** control panel:

7. Set **Arc Direction** to **CCW.**
8. Set **Radius** to **4.0".**

**FIGURE 7.5**
The tangent element start and end points of the 5" radius arc

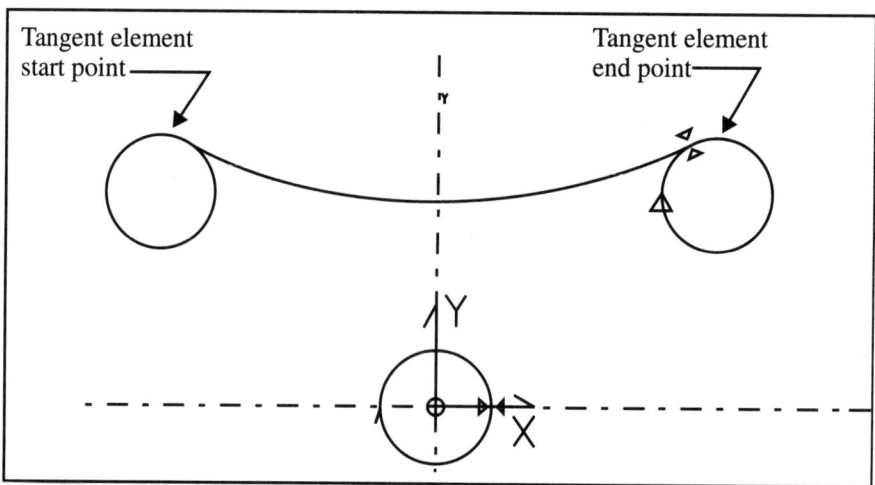

Chapter 7  SmartCAM Tutorial 6

You will again build this arc using the **Tangent El** function of the **Arc** control panel.

9. With your mouse, highlight the **S:** (tangent element start point).
10. Choose the lower right quadrant of the upper right arc.
11. For **E:** (tangent element end point) choose the lower right quadrant of the circle that is positioned at 0,0.

Your process model should now match Figure 7.6. Additionally, Figure 7.6 will further explain the mouse pickpoints described in steps 10 and 11.

12. Choose **Insert** from the workbench.
13. Select **After** and **Element Sequence** from the tool list.
14. Choose the circle that is positioned at X 0.0, Y 0.0.
15. Choose **Geometry** from the workbench.
16. Choose the **Arc** tool from the tool list.
17. **Arc Direction** should again be set to **CCW**.
18. **Radius** should again be set to 4.0".
19. Choose **S:** and choose the lower left quadrant of the circle that is positioned at X0.0, Y0.0.
20. For **E:** choose the lower left quadrant of the upper left circle.

Your process model should now match Figure 7.7.

Trim out the .50" radius circles until you get a blended profile which matches Figure 7.8.

**Remember:** *When using Trim/Extend pick the portion of the arc you wish to keep. Also, by turning **Which Segments** on, SmartCAM will show all possible solutions to the trim function.*

Continue with the model by placing the lead in/lead out moves on the 4.0" arc on the left side of the profile.

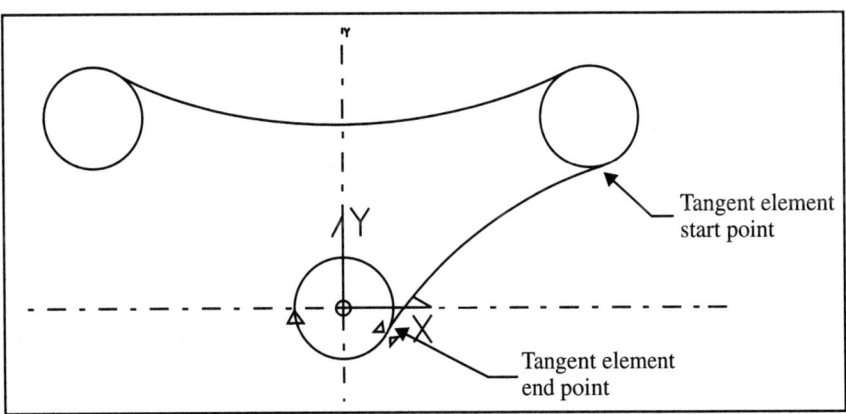

**FIGURE 7.6**
The tangent element start and end points of the first 4" radius arc

**FIGURE 7.7**
The tangent arcs of the exterior profile

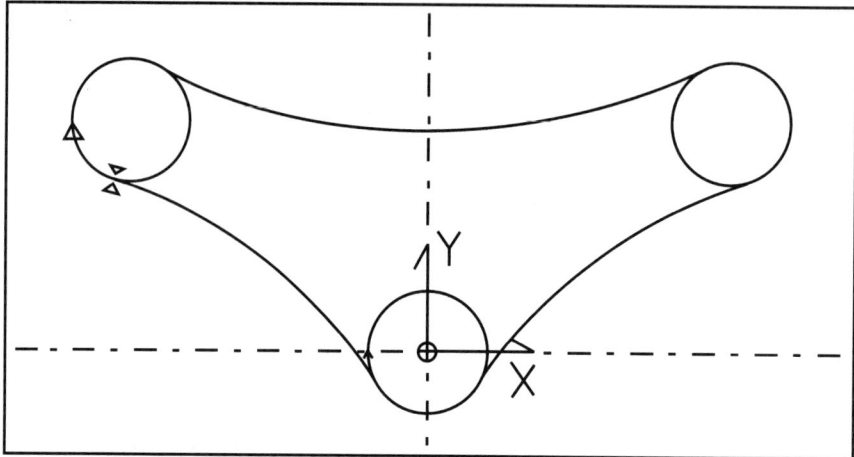

**FIGURE 7.8**
The completion of the trim function

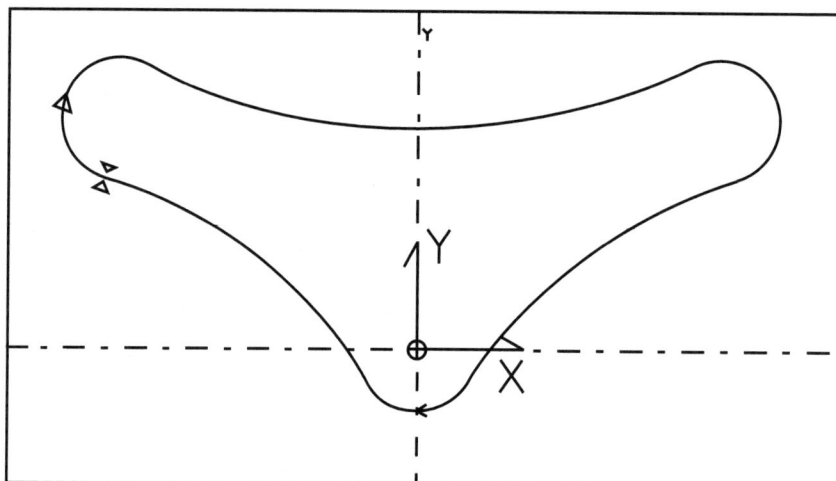

1. Choose **Geo Edit** from the workbench.
2. Select the **Split** tool from the tool list.
3. Inside the **Split** control panel, choose **Element Division** as the method to split the element.
4. Accept the default of .5 for the **% Length** input field.
5. Highlight **Select Split Element** and choose the 4.0" arc on the left side of the profile.
6. Choose **Lead in/Out** from the tool list.
7. Fill in the input fields of your **Lead in/Out** control panel to match those of Figure 7.9.

The addition of the lead in/lead out moves will complete your external finish profile. Your process model should now match Figure 7.10.

Before proceeding to the roughing profiles, verify the accuracy of the previously constructed geometry. Analyze the geometry in the database list and the different views of the graphics work area. Specific items to look for are geometry that is out of sequence and geometry that resides on different "Z" levels.

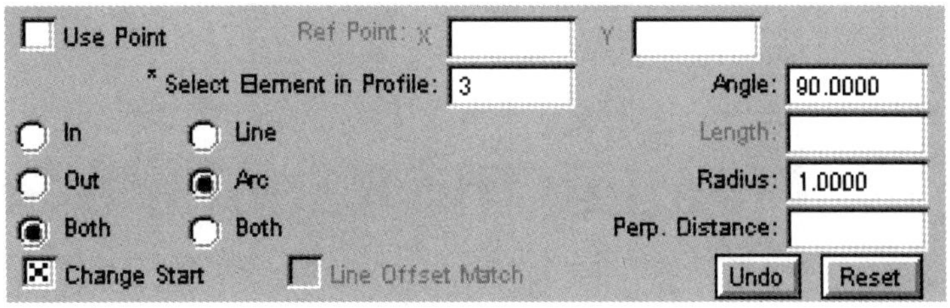

**FIGURE 7.9**
The input fields of the **Lead In/Out** control panel

**FIGURE 7.10**
The inserted lead in/lead out geometry

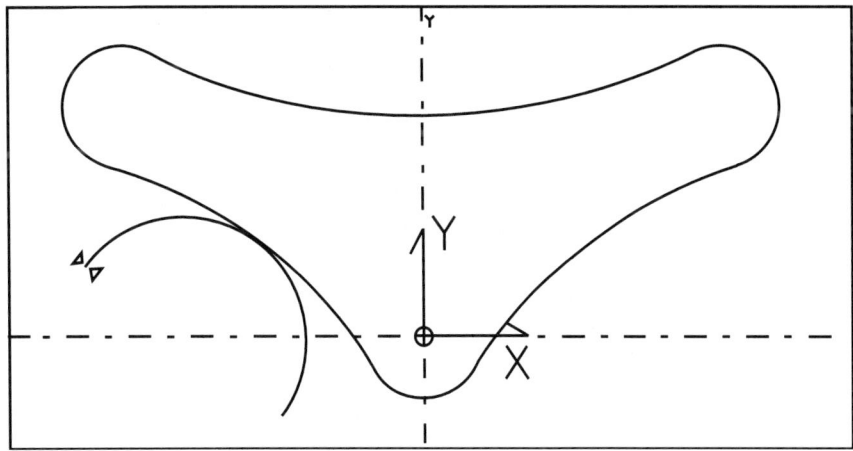

If the geometry is on different "Z" levels, it can be corrected by the following steps:

1. Choose **Group** from the workbench.
2. Choose the **Select All** button.
3. Choose **Edit** from the menu bar.
4. Choose the **Property Chg** option.
5. Select the **Toolpath** option.
6. Upon opening the **Toolpath Property Change** dialog box, make the required corrections.
7. Choose **Accept**.

If the constructed geometry is out of sequence, it can be corrected by the following steps:

1. Choose **Group** from the workbench.
2. Choose the **Select All** button.
3. Choose **Edit** from the menu bar.
4. Choose the **Order Path** toolbox from the list.
5. Select the **Chain** tool from the tool list.
6. Inside the **Chain** control panel, verify the Chain selection is on.
7. With your mouse, highlight the **Select an Element** in the profile input field.
8. Select the lead in arc.

Once the accuracy of the finish profile has been verified, proceed to the roughing geometry.

# Creating the Roughing Profiles

1. Choose **Group** from the workbench.
2. With your mouse, choose the **Select All** button.
3. Choose **Insert** from the workbench.
4. Choose the **Before**, **Step Sequence**, and **With Step** tools from the tool list.
5. Turn **Match Element** off.

Since the roughing step is not yet defined in the job planner, choose the **Add Step** button inside the **Insert** control panel to add a roughing step.

When the **Add Process Step** control panel opens, you will need to choose the operation category and type and the tool category and type as was done previously.

For **Op Category**, choose **Milling Operation**.

For **Op Type** choose **Rough Milling**.

For **Tool Category**, choose **Milling Tools**.

For **Tool Type**, choose **Endmill**.

Choose **Accept** and the job planner will open as normal.

6. Create a step that has the same properties as Table 7.2.

Once the properties for this step are entered properly, you can continue with the **Insert** control panel.

7. Fill in the input fields of your **Insert** control panel to match those of Figure 7.11.

To construct the first roughing profile:

1. Choose **Geometry** from the workbench.
2. Select the **Wall Offset** tool from the tool list.

Inside the Wall Offset control panel:

**TABLE 7.2**

| Step # | Tool # | Type | Diameter | Speed | Feed |
|---|---|---|---|---|---|
| 2 | 2 | End Mill (2 flute roughing) | 3/4" | 600 SFPM | 40 IPM |

**FIGURE 7.11**
The input fields of the Insert control panel

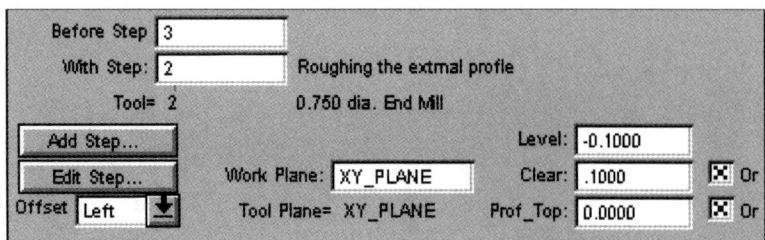

3. **Wall Side** should be set to left.
4. **Distance** should be set to .010".
5. Choose **Group Wall**.

At this point, your process model should match Figure 7.12. Continue the construction of the external roughing profiles.

1. Choose **Group** from the workbench.
2. With your mouse, select the **New Group** button.
3. Choose the **Step** option.
4. Choose step 2.
5. Choose **Edit** from the menu bar.
6. Select the **Transform** toolbox.
7. Choose the **Move** tool.

Inside the **Move** control panel:

8. Turn **Copy** on.
9. Enter 8 into the **Copies** input field.
10. With your mouse, choose the **From 0** button.
11. For **To Point**, input X 0.0, Y 0.0.

**FIGURE 7.12**
The exterior roughing and finishing profiles

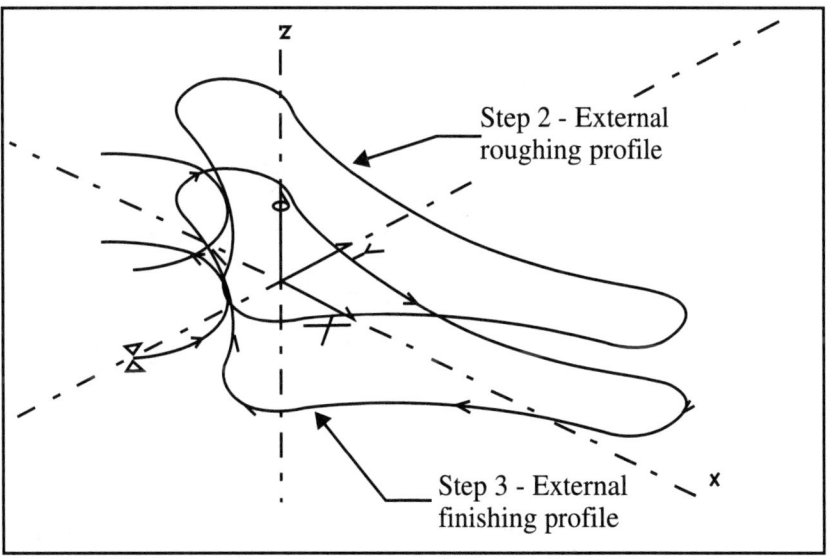

12. Stop when the Z input field is highlighted.

While the Z value is still highlighted:

13. Choose **Utility** from the menu bar.
14. Select the **Calculator** option.

For this example, the online calculator will be used to determine the correct Z value. When the calculator opens:

15. Input -.875/8 into the **Expression** input field.

    (-.875 is the depth of material which is left after the initial roughing pass of -.100. Eight is the value that was placed in the **Copies** input field).

16. Choose **Accept** and the result of the operation will be input into the Z input field.

However, if you will notice, the **Transform** operation did not occur. Simply repick the Z input field with your mouse and type the Enter key on your keyboard. The end result should be eight additional roughing passes on the external profile.

This will complete the roughing and finishing of the external profile. At this time, your process model should match Figure 7.13.

## Machining the Pocket

This project will require only a small amount of actual finish cuts to the pocket. The majority of stock will be removed with a pocket roughing routine. The only areas that will actually require finishing are a small portion of each of the three bosses. With that in mind, proceed directly to steps to create a pocket roughing routine.

To do a pocket roughing routine, you will have to construct a material boundary. This will be done with a layer since you do not actually want to machine the boundary.

**FIGURE 7.13**
The completion of the exterior roughing profiles

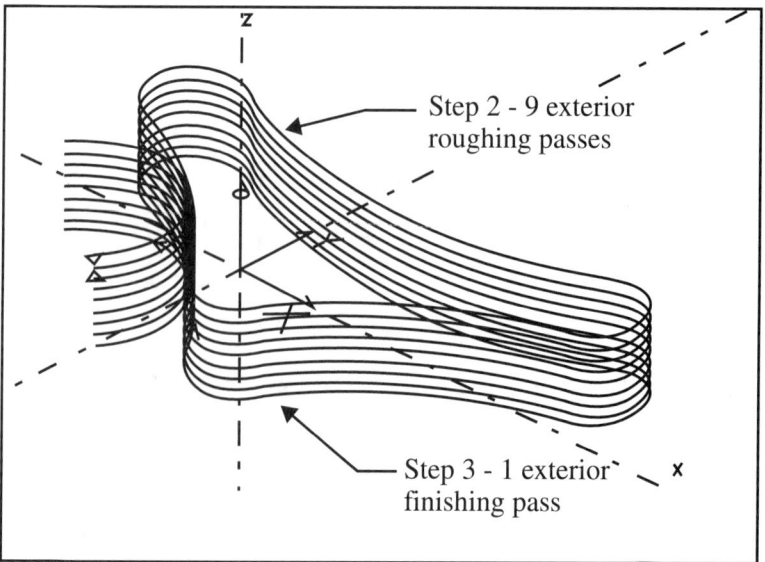

**FIGURE 7.14**
The input fields of the **Insert** control panel

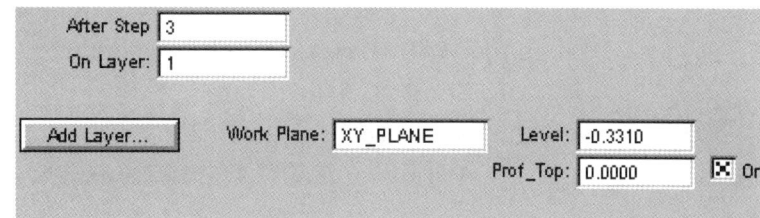

1. Choose **Insert** from the workbench.
2. Select the **After**, **Step Sequence**, and **On Layer** options from the tool list.
3. Fill in the input fields of your **Insert** control panel to match those of Figure 7.14.

   You need to create the material boundary at the same depth as the pocket; therefore, the **Level** is set at -.331". As usual, the **Prof_Top** is set to 0.000.

As discussed previously, when an element is defined as having a certain Z level and a profile top setting of zero, SmartCAM interprets the element as having a material thickness. As in this example, even though you will only see a thin profile, the fact that each element in that profile has a profile top of 0.00 and a Z level of -.331" actually gives that profile a material thickness. To fully demonstrate this situation:

1. Choose **Utility** from the menu bar.
2. Select the **Display Modes** option.
3. Inside the **Display Modes** dialog box, turn the **Thickness** selector switch to the on position (an x in the box indicates the on condition).

All elements will now reveal their "thickness" relative to their Z level and their profile top. To remove the appearance of the material thickness, simply turn the **Thickness** selector off. For the purposes of this tutorial, the **Thickness** selector switch is off.

Continue with your process model:

1. Choose **Group** from the menu bar.
2. With your mouse, select the **New Group** button.
3. Select the **Step** tool from the tool list.
4. Choose step 3, your finishing step.

Now, remove the lead in/lead out arcs from the active group. You will not need them to define your pocket material boundary.

5. Choose **Remove** and **Element** and pick the lead in arc and the lead out arc.
6. Choose **Create** from the menu bar.
7. Select the **Geometry** tool box.

8. Choose the **Wall Offset** tool from the list.

When the **Wall Offset** control panel opens:

9. Set **Wall Side** to left.
10. **Distance** is set to .375".
11. With your mouse, choose **Group Wall**.

You should now have a profile that defines the pocket, in addition to the external roughing and finishing passes.

Figure 7.15 shows the external finishing profile and the profile that defines the pocket. All other profiles have been masked to clarify the figure.

The next step in the construction of the pocket is to place the three bosses in their proper positions.

1. Obtain a full-scale, top view of your process model.
2. Choose **Insert** from the workbench.
3. Select the **After**, **Layer Sequence**, and **On Layer** tools from the tool list.

When the **Insert** control panel opens, input the following:

4. **After Layer** 1.
5. **On Layer** 1.
6. **Level** is set to -.331.
7. **Profile Top** is set to 0.000.
8. Choose **Geometry** from the workbench. If **Geometry** is not on the workbench, it can be found under the **Create** menu selection of the menu bar.
9. Choose the **Arc** tool.

Create three clockwise arcs, each with a radius of .500" at:

X 0.0, Y 0.0 ; X 2.5, Y 1.875 ; and X-2.5, Y1.875 respectively.

At this time, your process model should match Figure 7.16 All steps have been masked to clarify the figure.

**FIGURE 7.15**
The external finishing profile and the pocket boundary profile

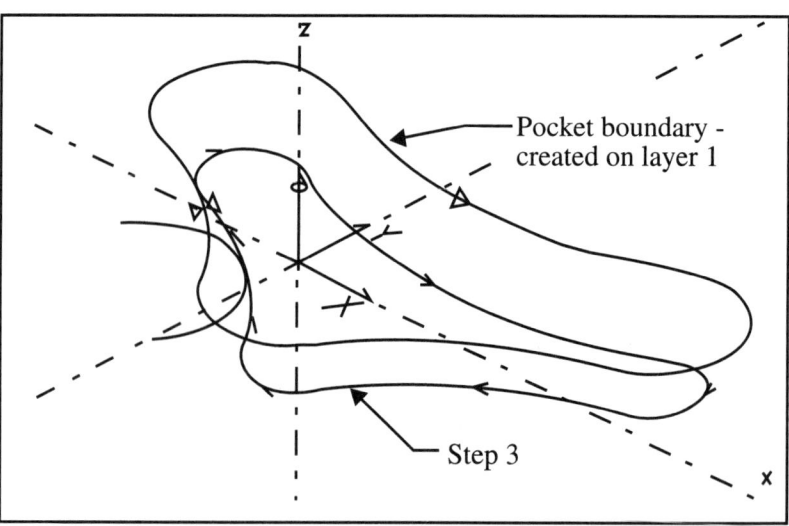

**FIGURE 7.16**
The 1″ diameter bosses

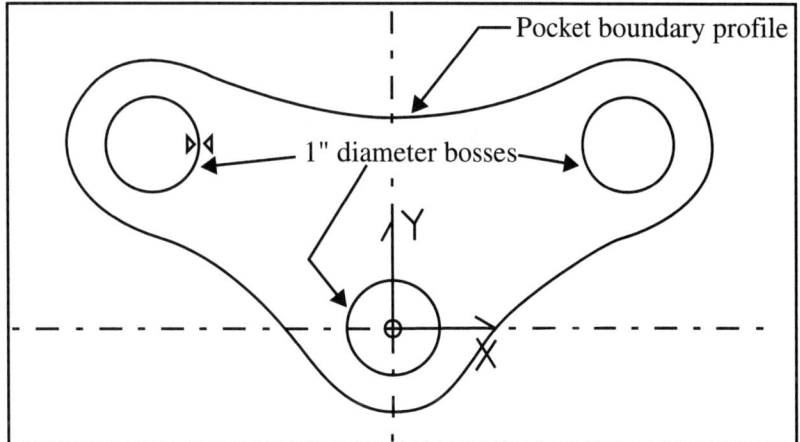

Continue by grouping the three arcs that were just constructed.

1. Choose **Group** from the workbench.
2. Select the **New Group** button.
3. Choose the **Element** option from the list.
4. With your mouse, select the three arcs from the graphics work area.

You are now ready to rough out the material that has been defined by the profiles in Figure 7.16.

1. Choose **Insert** from the workbench.
2. Select the **After**, **Step Sequence**, and **With Step** options from the tool list.

   Match Element should be off for this process.

Your pocket, however, will be roughed with a step that has not yet been defined. Therefore,

3. Choose **Add Step** from within the **Insert** control panel.
4. For **Op Category**, choose Milling Operation.
5. For **Op Type**, choose Rough Milling.
6. For **Tool Category**, choose Milling Tool.
7. For **Tool Type**, choose End Mill.
8. Choose the **Accept** button.
9. When the Job Planner opens, input Process Step 4.
10. Open the **Tool** page of the Job Planner, if it is not already open.
11. Since you will be using a predefined tool, tool 2, to rough your pocket, pick **Choose Tool** at the bottom of the tool page.
12. Highlight tool 2 from the tool list and pick **Use** at the bottom of the menu page.

The values from that predefined tool will be inserted into your current step.

13. Open the **Operation** page of the Job Planner.

**FIGURE 7.17**
The input fields of the **Insert** control panel

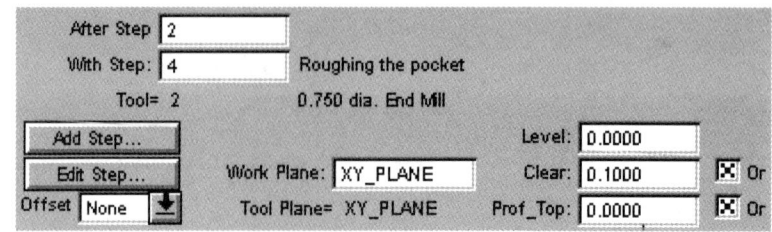

14. Input a cutting speed of 600 SFPM and a feed rate of 25 IPM.
15. Choose the **Accept** button.

Continue by inserting values in the input fields of your **Insert** control panel to match those of Figure 7.17.

16. Choose **Process** from the menu bar.
17. Select the **Rough** toolbox.
18. Choose the **Pocket** tool.
19. Fill in the input fields of your **Pocket** control panel to match those of Figure 7.18.

> For this example, make sure that the **Group Island** selector switch is on. The bosses will not be avoided if this switch is not on. (Actually, failure to avoid an island may come from several causes, one of which is the Z level of the island.) The island must be at the proper Z level, in this case -.331, and it must have a profile top of 0.00. However, there will be times when you will have to progressively shift the profile top in the positive direction until you arrive at a proper avoidance solution. In addition, each island must also be part of the active group.)
>
> The **Outside Boundary** is intentionally left blank for this figure. The **Outside Boundary** must be the profile that you created on the layer.
>
> All other input fields are fairly standard and may be adjusted according to your situation.

20. Once these input fields are correctly filled in, select the **Go** button with your mouse.

At this point, you should have a pocket roughing routine that resembles Figure 7.19. All external passes have been removed for clarity.

**FIGURE 7.18**
The input fields of the Pocket control panel

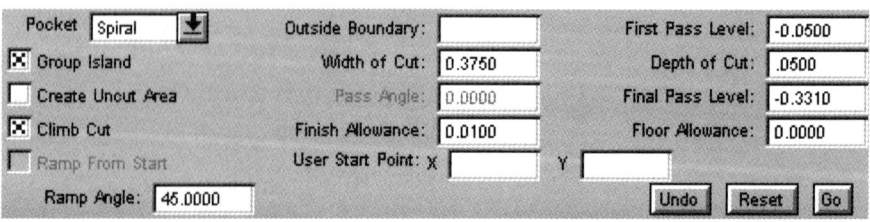

Chapter 7 SmartCAM Tutorial 6

**FIGURE 7.19**
The completed pocket roughing geometry

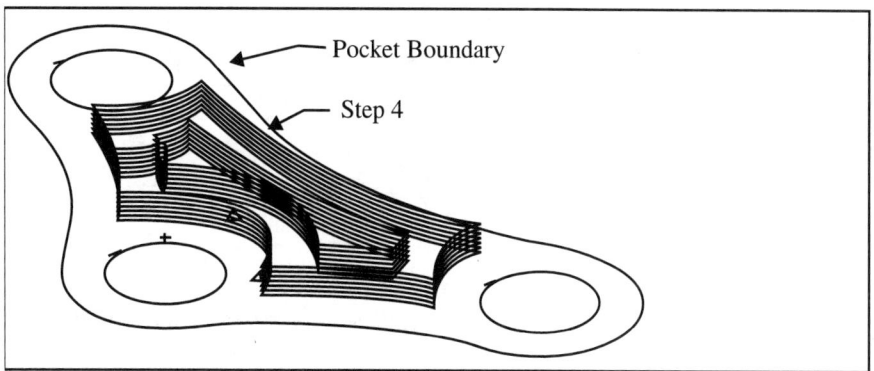

A close inspection of the pocket roughing cycle will reveal areas in the corners, close to the bosses, which were left unmachined. The 3/4" diameter endmill was simply too large to fit completely in these corner areas without invading the material boundary. In order to remove the excess material in these corners, you must now rough and then finish the inside portions of the island profiles. You will not need to machine the outer edges of these profiles as they were previously machined with the external roughing and finishing processes. You will limit your machining to the 180 degree arc segments that face the innermost portion of the pocket.

1. Choose **Insert** from the workbench.
2. Select the **After**, **Step Sequence**, and **With Step** tools from the tool list.
3. Fill in the input fields of your **Insert** control panel to match those of Figure 7.20.
4. Choose **Create** from the menu bar.
5. Select the **Geometry** toolbox.
6. Choose the **Arc** tool from the tool list.
7. Fill in the input fields of your **Arc** control panel to match those of Figure 7.21.

    If you will notice, the clockwise arc was placed at the origin. It started at 225.0 degrees and progressed around to 315.0 degrees. The next arc will have a center at X 2.5, Y1.875 and will have a starting angle of 300.0 degrees and ending angle of 110.0 degrees. The third arc will have a center of X-2.5, Y1.875

**FIGURE 7.20**
The input fields of the **Insert** control panel

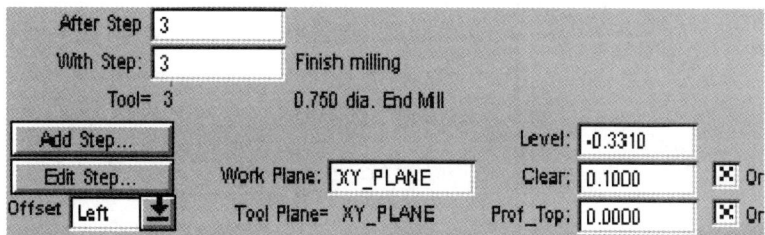

**FIGURE 7.21**
The input fields of the **Arc** control panel

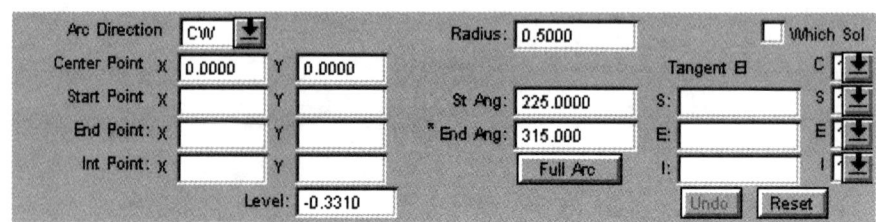

and will have a starting angle of 75.0 degrees and an ending angle of 240.0 degrees.

Before proceeding, construct the three arcs that were previously described.

You will now want to complete your arcs by adding a lead in/lead out move. For this example, both a lead in and a lead out arc of an angle of 90.0 degrees and a radius of 3/4" were chosen.

At this time, your process model should match Figure 7.22. (All roughing passes have been removed for clarity.)

Continue building your process model by placing the roughing profiles around the finish profiles you just created.

1. Choose **Group** from the workbench.
2. Select the **New Group** button with your mouse.
3. Select the **Profile** tool in the tool list.
4. With your mouse, select each profile that outlines the bosses in the graphics work area.
5. Choose **Insert** from the workbench.
6. Select the **After**, **Step Sequence**, and **With Step** options of the tool list.
7. Fill in the input fields of your **Insert** control panel to match those of Figure 7.23.

Continue with your model:

**FIGURE 7.22**
The lead in/lead out arcs of the bosses

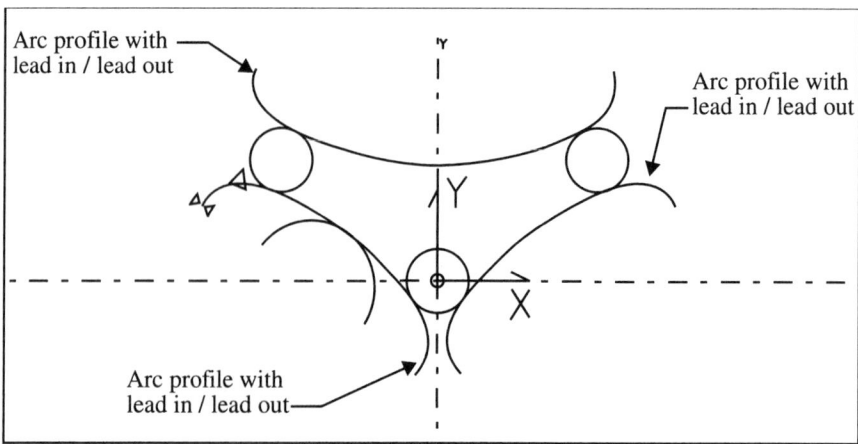

**FIGURE 7.23**
The input fields of the **Insert** control panel

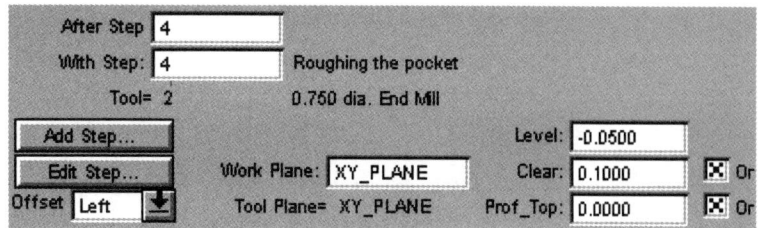

1. Choose **Geometry** from the workbench.
2. Select the **Wall Offset** tool.

Inside the **Wall Offset** control panel:

3. **Wall Side** should be set to left.
4. **Distance** should be set to .010".
5. With your mouse, choose the **Group Wall** button.

At this time, your model should show a roughing pass at the top of each island as shown in Figure 7.24.

Continue building the island roughing passes:

1. Choose **Group** from the workbench.
2. Select the **New Group** button.
3. Choose the **Profile** tool from the tool list.
4. Choose the three roughing profiles that were just created.
5. Choose **Edit** from the menu bar.
6. Select the **Transform** toolbox.
7. Choose the **Move** tool from the tool list.

**FIGURE 7.24**
The first roughing pass for island profile

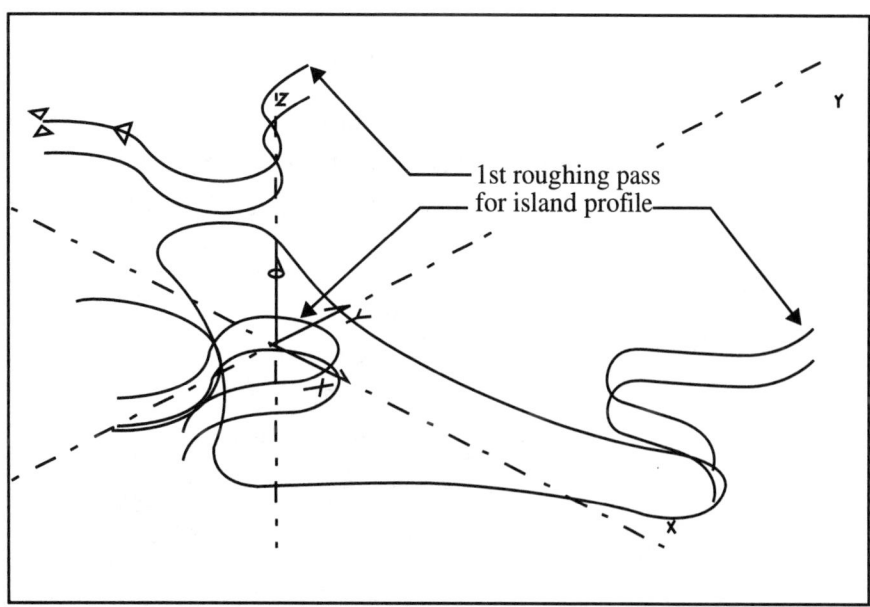

8. Turn **Copy** on and set **Copies** to 5.
9. With your mouse, select the **From 0** button.
10. For **To Point**, input 0.00 for X and Y.

For Z, while Z is highlighted:

11. Choose **Utility** from the menu bar.
12. Select the **Calculator** option.
13. For **Expression** enter the following: -(.331-.05)/5.

    The .331 value is the depth of the finish pass. The .05 is the material which was removed on the first roughing pass. Five is the number of copies. The mathematical formula will result in a value of -.0562″.

14. Choose **Accept** and the result will be transferred to the Z input field. Simply reclick on the Z input field and type the Enter key on your keyboard.

At this time, all roughing passes should be complete.

Your model should now match Figure 7.25. (The workplane indicator and axis indicators have been removed for clarity.)

# Alternate Pocket Roughing Routine

The pocket roughing routine that was described on the previous pages will work fine for most applications. However, as shown by **Show Path**, the endmill moves to the center of the part and plunges to depth. This is very hard on the cutting tools and on the machine itself. An alternate method of machining the pocket may be more desirable, depending on your application.

Build the pocket material boundary on a layer exactly as was described previously. You will, however, need to modify the boundary to provide a location for the cutting tool to plunge to depth. This location will be slightly off the material but inside the boundary.

To modify the boundary, follow these steps:

1. Choose **Edit** from the menu bar.
2. Select the **Geo Edit** toolbox.
3. Select the **Split** tool from the tool list.

**FIGURE 7.25**
The completion of the island roughing profiles

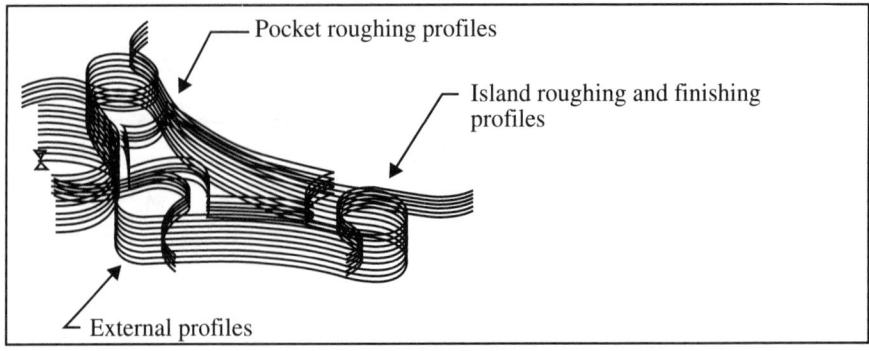

**FIGURE 7.26**
The split point of the pocket boundary profile

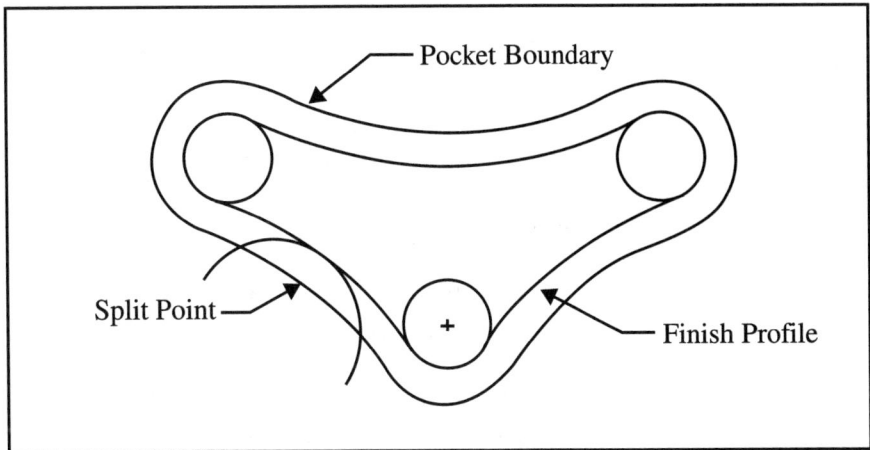

Split the boundary at the point shown in Figure 7.26. When the **Split** control panel opens:

1. Choose a **Gap Width** of .75″.
2. Select **Element Division** as the split method.
3. Accept .5 as **% Length**.
4. Choose the arc element as shown in Figure 7.26.
5. Insert a clockwise arc with a .375″ radius.
6. For the starting point, choose the lower end of the arc that was just split.
7. For the ending point, choose the upper end of the arc that was just split.

Upon completion of these steps, your process model should match Figure 7.27.

This arc that was inserted into your profile will provide clearance to begin your pocket roughing routine. However, you must still manipulate SmartCAM to make the roughing routine start at the desired location.

**FIGURE 7.27**
The inserted arc

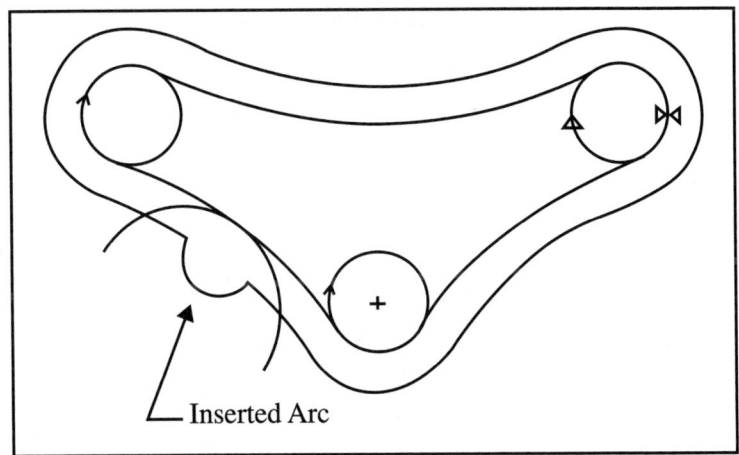

There are two major factors that control the starting point of the pocket roughing routine. One is the **Spiral Parameters** control panel that is opened from within the **Rough** toolbox. When this control panel is opened, a selector switch is available that will offer the option of cutting from the inside out (the switch is on for this option) or cutting from the outside in (the switch is off for this option). For the purposes of this example, the selector switch should be turned off (no x in the box).

The second factor that controls the starting point of the pocket roughing routine is the starting element of the material boundary. The pocket roughing routine starting position will be located close to the material boundary starting point. Therefore, you need to resequence your material boundary profile so that the first element in your profile is the .375" radius arc which you just inserted into your material boundary profile.

To do so:

1. Choose **Edit** from the menu bar.
2. Select the **Order Path** toolbox.
3. Choose the **Chain** tool from the tool list.
4. Inside the **Chain** control panel, choose the **Chain** option.
5. With your mouse, highlight **Select an Element in profile** and choose the arc that was inserted into your profile.

If you recall from previous examples, the purpose of **Chain** is to change nonsequential, connected elements into a contiguous profile. The element that was selected is also resequenced in the database list and becomes the starting element of the profile.

You are now ready to rough the pocket. Make sure the islands are the active group.

1. Choose **Insert** from the workbench.
2. Select the **After**, **Step Sequence**, and **With Step** options from the tool list.
3. Fill in the input fields of your **Insert** control panel to match those of Figure 7.28.
4. Choose **Process** from the menu bar.
5. Select the **Rough** toolbox.
6. Choose the **Spiral Parms** button from the tool list.

**FIGURE 7.28**
The input fields of the **Insert** control panel

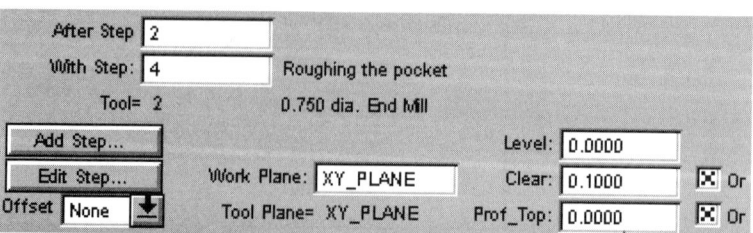

**FIGURE 7.29**
The input fields of the **Pocket** control panel

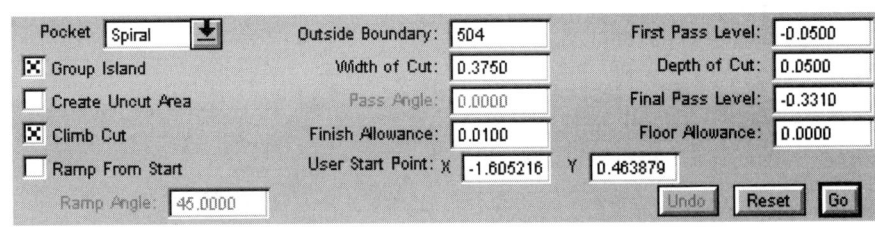

**FIGURE 7.30**
The user start point error message

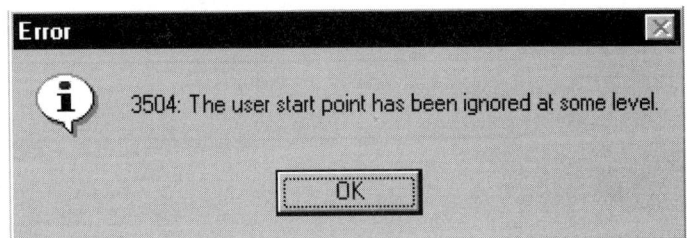

Inside the **Spiral Parms** dialog box,

7. Turn the **Cut Inside Out** selector switch off.
8. Choose **Pocket** and fill in the input fields of your **Pocket** control panel to match those of Figure 7.29.

This control panel is only slightly different from the previous pocket roughing routine example. For **Outside Boundary**, select the .375" radius arc which you inserted into your material boundary profile. Additionally, this example will make use of the **User Start Point**. The **User Start Point** allows you to choose the starting point of the roughing profiles. Highlight **User Start Point** and choose the center point of the .375" radius arc. It is very important, however, that you turn off the **Cut Inside Out** option inside the **Spiral Parms** control panel.

9. When you complete the input fields choose the **Go** button.

As your roughing profiles are created, you will probably see the error message shown in Figure 7.30.

This is just a warning from SmartCAM notifying you that your user start point values could not be adhered to for all Z levels. It will not severely affect your roughing passes. Simply pick the **OK** button and the message will disappear.

At this point, the roughing passes of your process model should match Figure 7.31.

A close-up view of the starting position of the pocket roughing routine (Figure 7.32) will illustrate the correct outcome of your pocket roughing routine.

## Spot Drilling and Drilling the Holes

Continue with your model by placing the spot drill holes in their proper locations.

**FIGURE 7.31**
The completion of the alternate pocket roughing routine

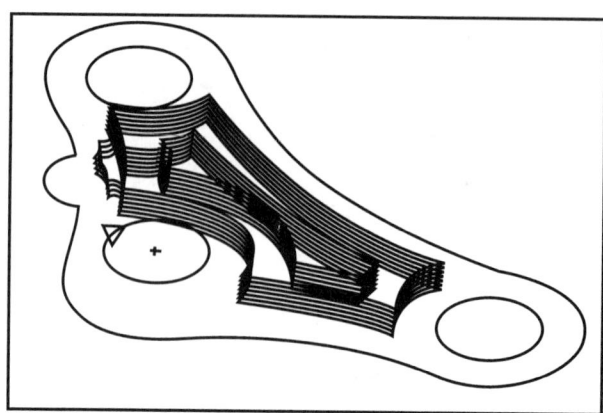

**FIGURE 7.32**
The starting position of the pocket roughing routine

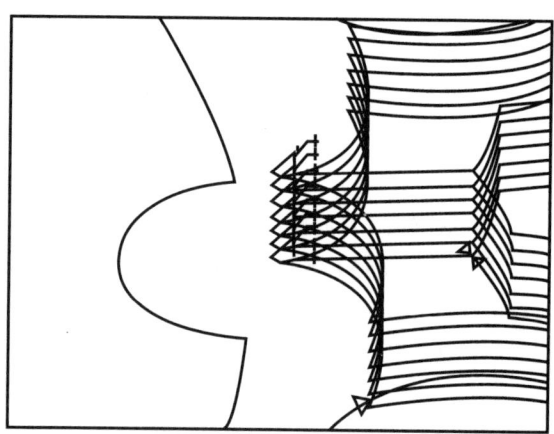

1. Choose **Insert** from the workbench.
2. Select the **After**, **Step**, and **With Step** options from the tool list.
3. When the **Insert** control panel opens choose **Add Step**.

It will be necessary to add another step as you have not yet defined a spot drill in your job planner. Inside the **Add Process Step** dialog box:

1. Choose **Hole Operation**.
2. Choose **Spot Drilling**.
3. Choose **Hole Tool.**
4. Choose **Spot Drill.**

When the job planner opens, input the information in Table 7.3.

Once the additional step is completed in the job planner, you will be automatically returned to the **Insert** control panel.

**TABLE 7.3**

| Step # | Tool # | Type | Diameter | Speeds | Feeds |
|---|---|---|---|---|---|
| 5 | 4 | Spot Drill | 5/16" | 600 SFPM | 12 IPM |

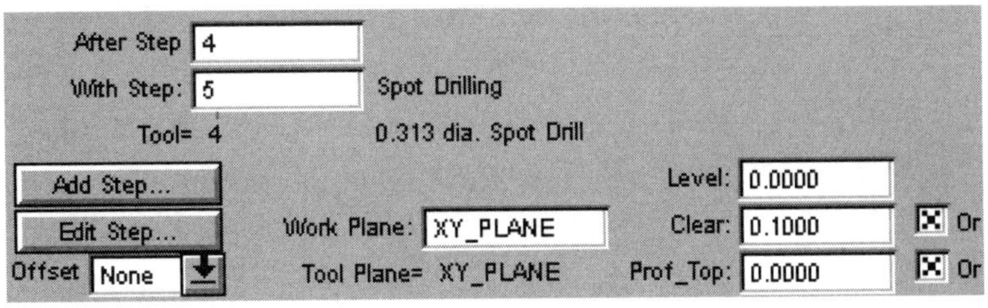

Chapter 7  SmartCAM Tutorial 6

**FIGURE 7.33**
The input fields of the **Insert** control panel

Fill in the input fields of your **Insert** control panel to match those of Figure 7.33.

1. Choose **Geometry** from the workbench.
2. Select the **Hole** tool from the tool list.
3. Create a hole with a Spot Dia of .3125" at a location of X0.0, Y0.0.
4. Choose **Insert** from the workbench.
5. Select the **After**, **Step Sequence**, and **With Step** tools from the tool list.
6. Inside the Insert control panel, choose the **Add Step** button.

Once you are inside the **Add Process Step** dialog box:

7. Choose **Hole Operation**.
8. Choose **Drilling**.
9. Choose **Hole Tool**.
10. Choose **Twist Drill.**

When the job planner opens, input the information in Table 7.4. Complete your **Insert** control panel by entering:

1. **After Step 5**, **With Step 6**. All other fields will be the same as the spot drill.
2. Choose the **Geometry** toolbox.
3. Select the **Hole** tool.
4. Enter a value of .975 for **Full Depth**.
5. Highlight the input field **Hole Point** and, with your mouse, pick the spot drill hole that was previously placed at the origin.

Your graphics work area and your database list should now reveal a spot drill hole and a drill hole.

You will simply **Transform**, **Copy** the holes to the other locations.

**TABLE 7.4**

| Step # | Tool # | Type | Diameter | Speeds | Feeds |
|---|---|---|---|---|---|
| 6 | 5 | Drill | 5/16" | 600 SFPM | 10 IPM |

1. Choose **Group** from the workbench.
2. Select the **New Group** button.
3. Choose the **Step** tool and choose steps 5 and 6.
4. Choose **Edit** from the menu bar.
5. Select the **Transform** toolbox.
6. Choose the **Move** tool from the tool list.

Inside the **Move** control panel:

7. Turn **Copy** on and set **Copies** to 1. Make sure the **Sort by Tools** selector switch is turned on.
8. Choose **From 0**.
9. For **To Point**, input X 2.5, Y 1.875, Z 0.0.
10. For the third hole choose **From 0**.
11. For **To Point**, input X -2.5, Y 1.875, Z0.0.

All holes should now be placed on your process model.

## Face Milling the Work

The next step in the construction of your process model is the face milling operation. Again, since this step is undefined, you must first define this step in your job planner.

1. Choose **Insert** from the workbench.
2. Select the **Before**, **Step Sequence**, and **With Step** tools from the tool list.
3. Choose the **Add Step** button from within the **Insert** control panel.

This step will be a face milling operation with a 3″ diameter face mill. Choose the appropriate selection for each category within the **Add Process Step** dialog box. Once the job planner has opened, input the proper values into each input field from Table 7.5.

Upon completion of your job operations planner, fill in the input fields of your **Insert** control panel to match those of Figure 7.34.

1. Obtain a full-scale, top view of your model.
2. Choose **Process** from the menu bar.
3. Select the **Rough** toolbox.
4. Choose the **Face** tool.

**TABLE 7.5**

| Step # | Tool # | Type | Diameter | Speeds | Feeds |
|---|---|---|---|---|---|
| 1 | 1 | Face Mill | 3″ | 600 SFPM | .010 IPT |

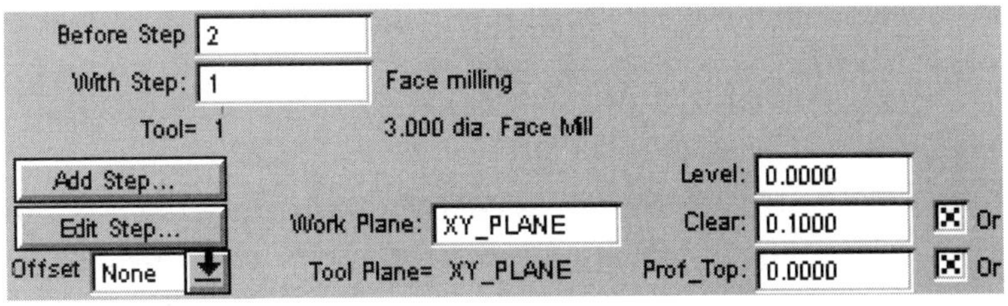

Chapter 7   SmartCAM Tutorial 6     161

**FIGURE 7.34**
The input fields of the **Insert** control panel

5. Fill in the input fields of your **Face** control panel to match those of Figure 7.35.

If you recall from the previous tutorials where a facemill was used, the first facemill passes were all constructed at a level of Z 0.0. In order to remove material, the G54 "Z" level was adjusted at the machine tool. This is also the way this example is shown. If, however, you prefer to adjust for a depth of cut within the **Face** roughing control panel, it is a simple matter of entering values for the **First Pass Level**, **Depth of Cut,** and **Final Pass Level** input fields. You must, however, adjust the other steps the appropriate amount to compensate for the material removed by this method.

If you analyze your facemill geometry by running Show Path or by choosing **Utility > Element Data**, you will notice several passes that are seemingly unnecessary. SmartCAM created passes which would properly allow a **Width of Cut** of 1.500". Occasionally you will simply need to delete the unnecessary passes that are created outside of the material boundary.

Choose **Edit > Geo Edit > Delete** and choose the unwanted geometry.

Once the unwanted geometry has been removed, you can proceed with copying the face mill passes down .015". For the sake of this example, assume that the material stock size is 1.00". You will remove .010" from the top and .015" from the bottom to arrive at the finish size of .975".

To copy the face mill passes:

1. Choose **Group** from the workbench.
2. Select the **New Group** button.
3. Choose the **Step** tool and choose Step 1.
4. Choose **Edit** from the menu bar.
5. Select the **Transform** toolbox.
6. Choose the **Move** tool.

When the **Move** control panel opens:

7. Turn **Copy** on and set **Copies** to 1.

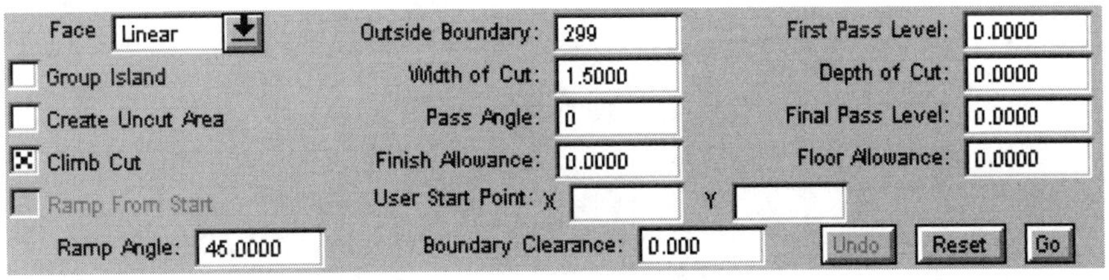

**FIGURE 7.35**
The input fields of the **Face** control panel

8. Choose the **From 0** button.
9. For **To Point** input X0.0, Y0.0, Z-.015″.

To accomplish your goal of face milling both sides of the workpiece, you will need to flip the work after the facing passes which were created at Z 0.0″. This will require you to insert a **User Event** between the two facemilling passes exactly as was done in the previous tutorial.

To do so:

1. Choose **Insert** from the workbench.
2. Select the **After**, **Element Sequence**, and **With Step** options from the tool list.
3. Fill in the input fields of your **Insert** control panel to match those of Figure 7.36.

    Element 3 was the last pass of the facemill geometry that was created at Z0.0.

    **With Step** must be active as the **User Event** must have toolpath properties in order for the code generator to output the user event into the code file. The remaining input fields were either left blank or the default was accepted as they are not critical for a **User Event**.

4. Choose **Create** from the menu bar.
5. Select the **User Elemts** toolbox.
6. Choose the **User Event** tool.
7. When the **User Event** control panel opens, input "G28 M00."
8. Choose a **Location Point** that is within the current work area. The value is not critical, however, a valid number must be entered.
9. Choose **Go** and the User Event will be placed into your database list.

Recall from the previous tutorial that the M00 will cut off all machine movement, as well as the spindle speed and the coolant. Continue placing User Events into your database list to resume all functions that were stopped by the M00.

Recall also from the previous tutorial, that in order for your code generator to output an X and a Y value after the user event, it was necessary to insert a Point/Rapid move into your database list. Because the start point

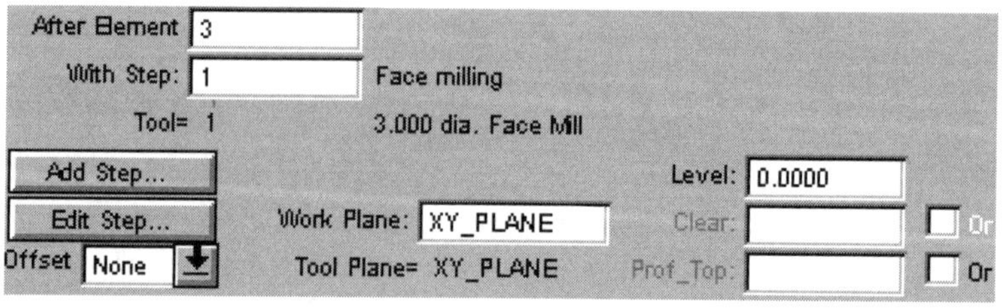

**FIGURE 7.36**
The input fields of the **Insert** control panel

Chapter 7 SmartCAM Tutorial 6

coordinates of the line after the user event are different from the end point coordinates of the line immediately before the user event, a Point/Rapid move is not necessary. Since the values are different, both an X and a Y coordinate will automatically be output into the code file. In the previous tutorial, only the X coordinate of the line changed. That resulted in only an X value being output into your code file.

## Constructing Additional Workplanes

The last major operation of this tutorial is the construction of two additional workplanes. A G55 workplane will be constructed so that you can mill the entire profile and one of the pockets. A G56 workplane will be constructed so that the second pocket can be milled. In actual machining, only two fixtures will be needed. The part will simply be flipped at each station. However, in SmartCAM, you will need three workplanes to more clearly view your process model. At the machine tool, the G55 and the G56 workplanes will simply have the same coordinate values.

1. Choose **Workplane** from the menu bar.
2. Select the **Define Plane** option from the list.
3. Fill in the input fields of your **Define Plane** control panel to match those of Figure 7.37.

If you recall from the previous tutorial, the **workplane Name** and the **Tool Plane** name must match so that proper code can be generated. Recall also that all values are based on the active workplane, which in this example is the G54 workplane. Once these values are accepted, the G55 workplane becomes the active workplane.

Continue building the G56 workplane exactly as you did the G55 workplane. The only difference will be the **Plane Name** and the **Tool Plane**. Input a G56 for these two input fields.

The next step is to move the appropriate geometry to the proper workplane.

1. Choose **Group** from the workbench.
2. Select the **New Group** button.
3. Choose the **Step** tool and choose steps 2, 3, and 4.

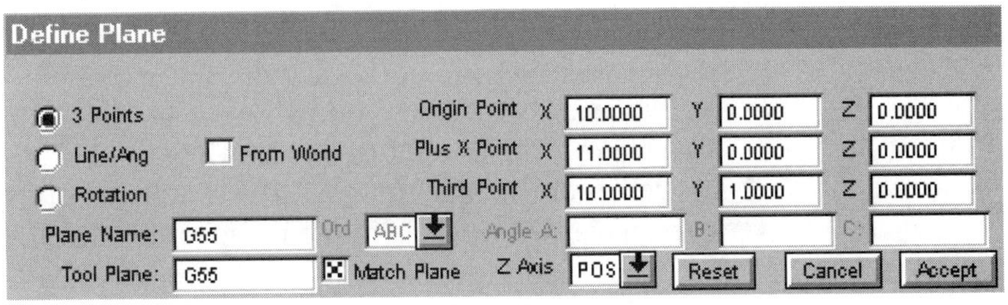

**FIGURE 7.37**
The input fields of the **Define Plane** control panel

## Chapter 7 SmartCAM Tutorial 6

4. Choose **Edit** from the menu bar.
5. Choose the **Transform** toolbox.
6. Choose the **Move** tool from the tool list.
7. Select the **From 0** button.
8. With your mouse, highlight **Destination Plane**.
9. Choose G55 from the workplane listing.

The steps should move to the G55 workplane.

10. Next, choose **Group > New Group > Step**.
11. Choose steps 3 and 4.

The finishing step, step 3, is associated with finishing passes on both the external geometry and the bosses. You only want the geometry that is associated with the bosses to be copied over to the G56 workplane.

12. Choose **Remove > Profile** and choose the external finish profile (you may need to window in and choose the profile very carefully).
13. Return to the **Transform > Move** control panel.
14. Turn **Copy** on and set **Copies** to 1.
15. Choose **Destination Plane** and choose G56 from the workplane listing.

The pocket roughing routine and the geometry that roughs and finishes the bosses should be copied to the G56 workplane.

Once all geometry is located on the proper workplanes, you can insert the appropriate **User Events** that are necessary for each workplane.

1. Choose **Insert** from the workbench.
2. Select **Before Step** and **With Step** from the tool list.
3. Input before step 2, with step 2.
4. Choose **Workplane** from within the **Insert** control panel and input G55.
5. Accept the default for all other input fields.
6. Choose **Create > User Elmts > User Event**.

Insert the following **User Events** into your database:

G28 M00 (This will move the cutting tool to a safe position so you can flip the workpiece and stop the movement of the machine.)

S3000 M03 (Spindle on - clockwise)

M08 (Coolant on)

G00 G55 (Set coordinates to the G55 workplane)

This exact grouping of user elements will be inserted before the third workplane also. Again, you will need to make the decision whether to insert two G55 workplanes or to insert a G55 and a G56 workplane. In the code file, as well as at the machine tool, it is a matter of preference. You may not, however, have two workplanes with the same name in SmartCAM.

Run the Show Path function of SmartCAM and very carefully watch the tool to see the order in which the steps are machined. The steps will, quite

Chapter 7 SmartCAM Tutorial 6

possibly, be out of order. If the steps are machined out of sequence, you will need to resequence your database prior to generating code.

To do so:

1. Choose **Group** > **New Group** > **Step** and choose steps 1, 5, and 6.
2. Change the group selection method to **Box** and box in the entire G55 workplane.
3. Next, box in the entire G56 workplane.
4. Choose **Sequence Move** from inside the **Group** control panel (make sure the **By Selection Order** selector switch is turned on). The steps will be resequenced in the database listing according to the order in which they were selected.

At this point, your sixth tutorial is complete and should match Figures 7.38, 7.39. and 7.40.

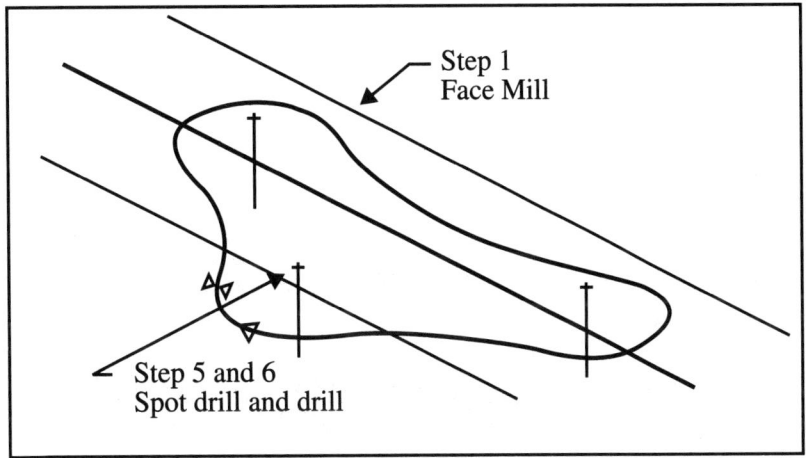

**FIGURE 7.38**
The G54(xy workplane) shown with face mill, spot drill, and drill steps; the material boundary is shown on layer 1

**FIGURE 7.39**
The G55 workplane shown with pocket milling and profile roughing and finishing milling operations

**FIGURE 7.40**
The G56 workplane shown with second pocket milling routine, profile roughing, and finishing for the bosses

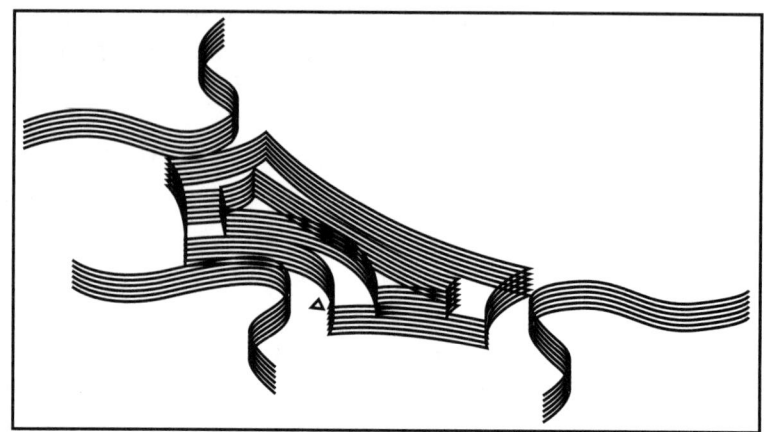

# CHAPTER 8

# SmartCAM Tutorial 7

## Idler Wheel Support

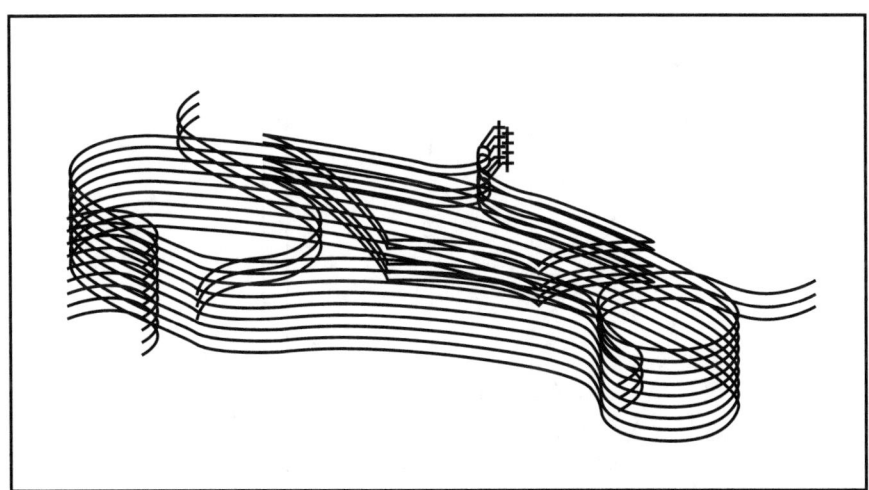

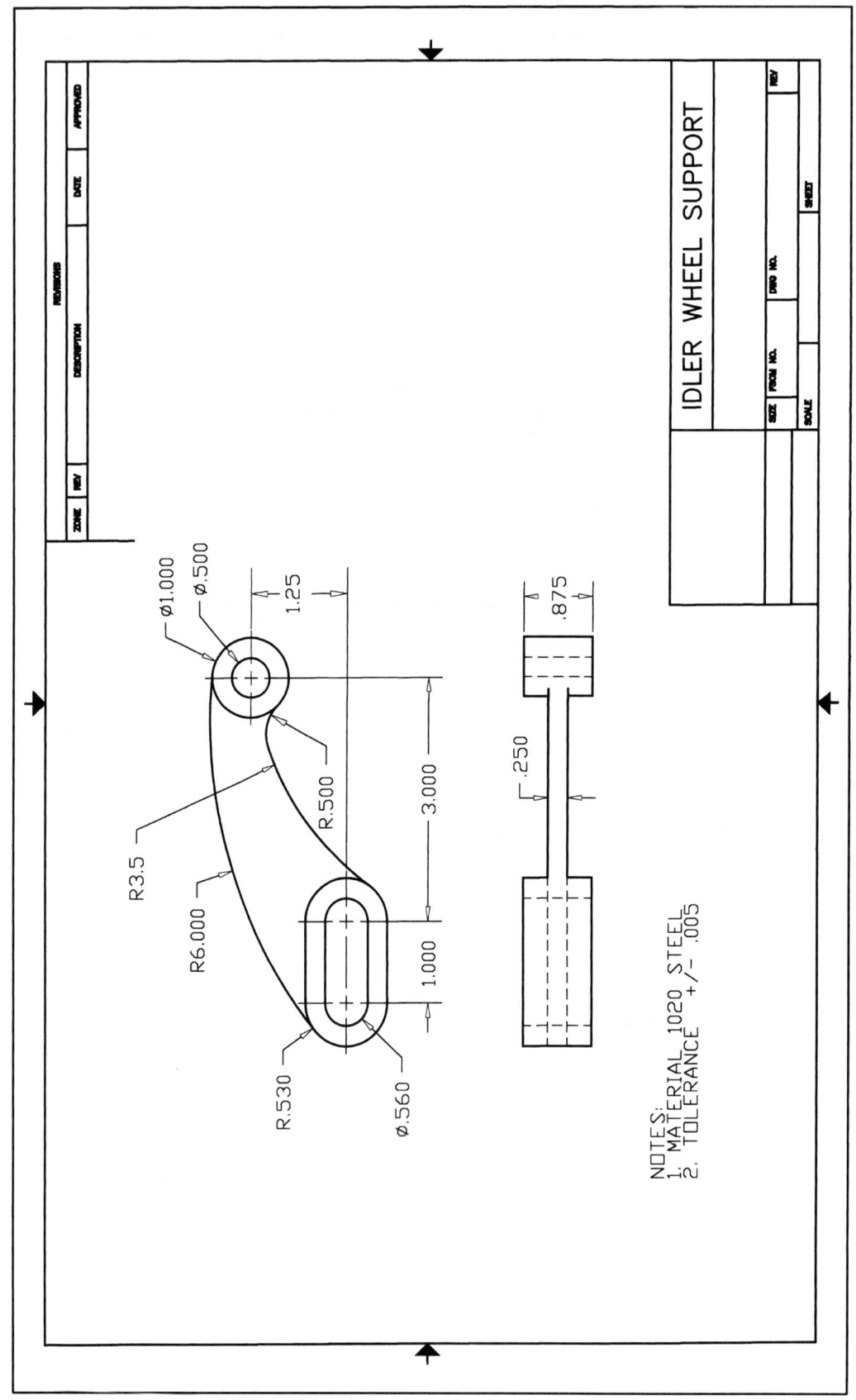

The blueprint for the Idler Wheel Support Tutorial

Chapter 8  SmartCAM Tutorial 7

Upon completion of this chapter, you should be able to:

- Read in and modify an existing job operations file.
- Construct more complex geometry consisting of tangent arcs and lines.
- Manipulate profiles using the Order Path function of SmartCAM.
- Be more proficient in the construction of multiple workplanes.
- Identify and solve the errors of out-of-sequence geometry.

# IDLER WHEEL SUPPORT

This last tutorial, the Idler Wheel Support, consists of geometry that is more complex in nature. The construction of the Idler Wheel Support will provide an opportunity to reinforce all previously learned skills. A conscious effort was made to introduce very few, if any, new skills.

This tutorial will use a slightly modified version of the job file that was used in Tutorial 5, The Four-Hole Frame. Read in this job file and make the modifications according to the following example.

To read in this job file:

1. Select **File**.
2. Select **Load Job File**.
3. Fill in the **Load Job File** dialog box with the appropriate file name.
4. Choose the **Accept** button.

If by chance the file was lost and you are unable to read in the correct job file, build a job operation file that contains the tools and steps in Table 8.1.

As always, begin the construction of your process model with the finish geometry. To do so, follow these steps:

1. From the workbench choose the **Insert** toolbox.
2. Choose the **After** and the **Element Sequence** modeling tools from the tool list.
3. Choose the **With Step** option.
4. Fill in the input fields of your **Insert** control panel to match Figure 8.1.
5. Choose **Geometry** from the workbench.
6. Choose **Arc** from the tool list.
7. Fill in the input fields of your **Arc** control panel to match Figure 8.2.
8. Continue with the **Arc** control panel to construct the next arc.
9. For the **Center Point** input fields enter X 1.00, Y 0.00.
10. Accept the default values for all other input fields.

**TABLE 8.1**

| Step # | Description | Tool # | Type | Diameter | Speeds | Feeds |
|---|---|---|---|---|---|---|
| 1 | Facing the blank material | 1 | Face Mill (Carbide) | 3" | 300 SFPM | .010 IPT |
| 2 | Roughing the profile | 2 | End Mill (2 flute roughing) | 5/8" | 80 SFPM | .004 IPT |
| 3 | Roughing the pocket | 2 | End Mill (2 flute roughing) | 5/8" | 80 SFPM | .003 IPT |
| 4 | Finishing the profile | 3 | End Mill (4 flute finishing) | 5/8" | 90 SFPM | .003 IPT |
| 5 | Roughing the slot | 4 | End Mill (2 flute roughing) | 1/2" | 80 SFPM | .002 IPT |
| 6 | Finishing the slot | 5 | End Mill (4 flute finishing) | 1/2" | 90 SFPM | .0015 IPT |
| 7 | Spot drilling | 6 | Spot Drill | 3/8" (90 deg) | 90 SFPM | .005 IPR |
| 8 | Drilling | 7 | Drill | 1/2" | 90 SFPM | .007 IPR |

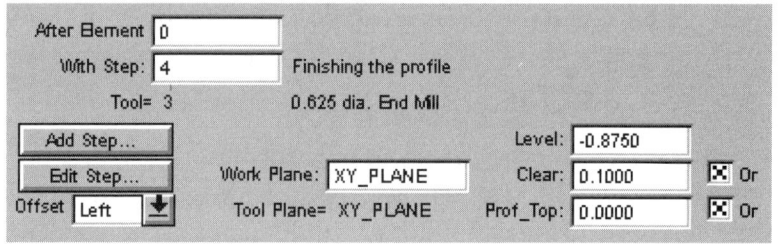

**FIGURE 8.1**
The input fields of the **Insert** control panel

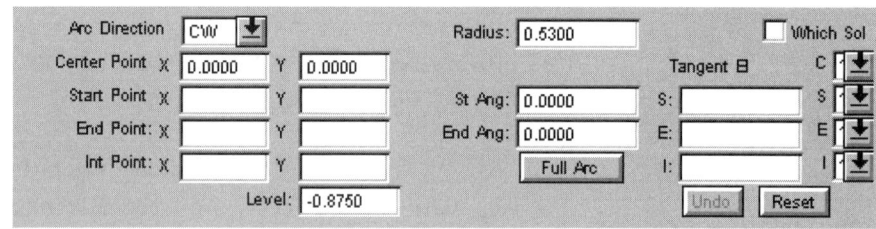

**FIGURE 8.2**
The input fields of the **Arc** control panel

You should now have two full arcs shown in your graphics work area. Continue by placing two lines tangent to the two arcs.

1. Choose the **Line** modeling tool from the tool list.

You will be using the **Tan Arc** method of creating the two lines.

2. Choose the **Start** input field.
3. Select the top of the left arc.
4. Select the top of the right arc to input the element number in the **End** input field.

**FIGURE 8.3**
The construction of the elongated boss profile

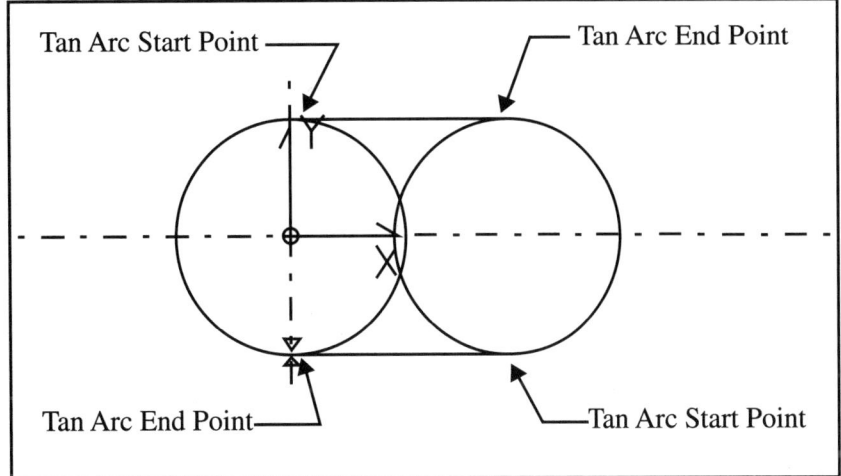

A line should appear which is tangent to the two arcs.

5. Highlight the **Start** input field and select the bottom of the right arc.
6. Select the bottom of the left arc to input the element number in the **End** input field.

At this time your process model should match Figure 8.3.
Continue with your process model by trimming the arcs.

1. Choose **Geo Edit** from the workbench.
2. Choose **Trim/Extend**.
3. For the **Select 1st Element** input field, choose the left arc on the left side of the vertical centerline of the arc (at approximately the 180 degree point).
4. For the **Select 2nd Element** input field, choose the bottom line.
5. Repeat step 3.
6. For the **Select 2nd Element** input field, choose the top line.

The left arc should now be trimmed as per the print.
To trim the right arc:

1. Verify that the **Select 1st Element** input field is highlighted. If it is not, select it at this time.
2. Select the right arc in the bottom right quadrant. (At approximately the 315 degree point).
3. For the **Select 2nd Element** input field, choose the bottom line.
4. Repeat steps 1 and 2.
5. For the **Select 2nd Element** input field, choose the upper line.

Your process model should now match Figure 8.4.

**FIGURE 8.4**
The trimmed arcs

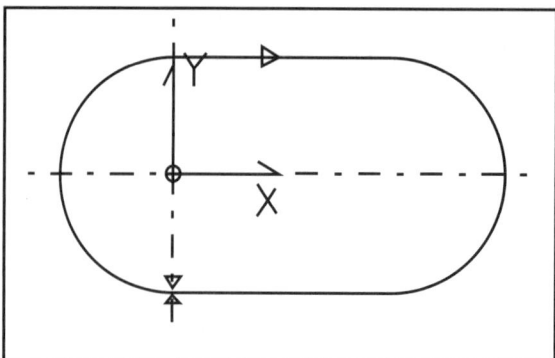

Continue with your process model by placing the 1.00″ diameter boss as shown in the blueprint.

1. Choose **Geometry** from the workbench.
2. Choose the **Arc** modeling tool from the tool list.
3. Fill in the input fields of your **Arc** control panel as per Figure 8.5.

After filling in the **Arc Direction**, **Radius**, **Center Point** and **Level** input fields, simply select the **Full Arc** button and the geometry will be correctly placed.

At this time, your process model should match Figure 8.6.

You will continue with the process model by constructing the 6.0″ radius arc. To correctly build it, you must first resequence the database.

**FIGURE 8.5**
The input fields of the **Arc** control panel

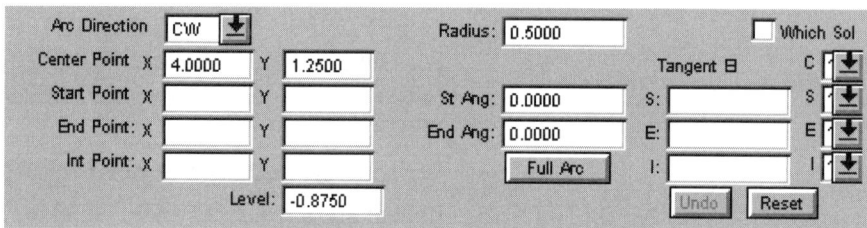

**FIGURE 8.6**
The completion of the elongated boss and 1″ diameter boss profiles

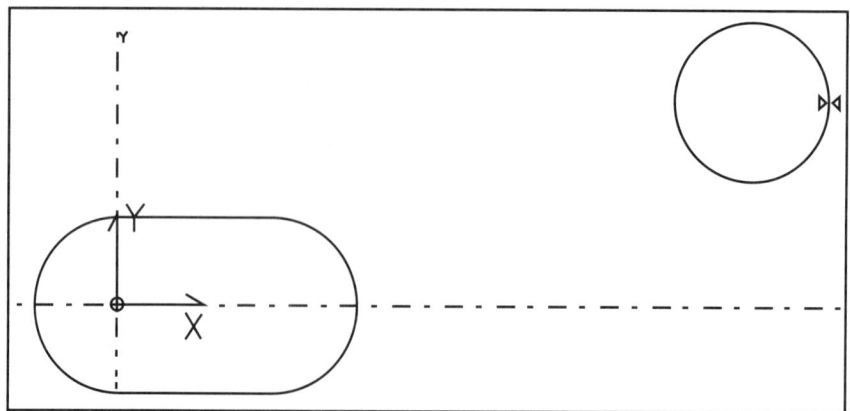

# Chapter 8  SmartCAM Tutorial 7

1. Choose **Insert** from the workbench.
2. Select **After**, **Element Sequence**, and **With Step** from the tool list.
3. Within the **Insert** control panel, input "1" for the **After Element** input field. This should be the arc on the bottom left (accept the default for all other input fields).
4. Choose **Geometry** from the workbench.
5. Choose **Arc**.

   **Arc Direction** will be clockwise.

   **Radius** will be 6.0.

To create the arc, you will again use the **Tangent El** function of the Arc control panel as was demonstrated in a previous tutorial. Input the values as shown in Figure 8.7.

The simplest way to input these values is to highlight each input field and then, from the graphics work area, select the tangent points with your mouse. Refer to Figure 8.8 for further clarification.

Your next objective is to construct the .50" radius arc as per the print.

1. Choose **Insert**.
2. Choose **After** and **Element Sequence.**
3. Turn on **Match Element**.
4. For the **After Element** input field, select the 1.0" diameter boss on the right of the model.

**FIGURE 8.7**
The **Tangent Element** function of the **Arc** control panel

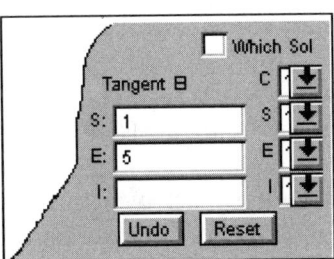

**FIGURE 8.8**
The Tangent Element start and end points of the 6" radius arc

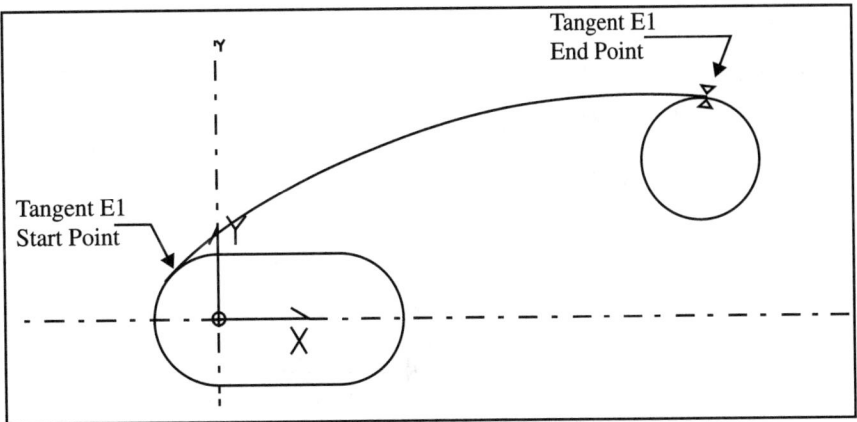

**FIGURE 8.9**
The construction of the .50" radius arc

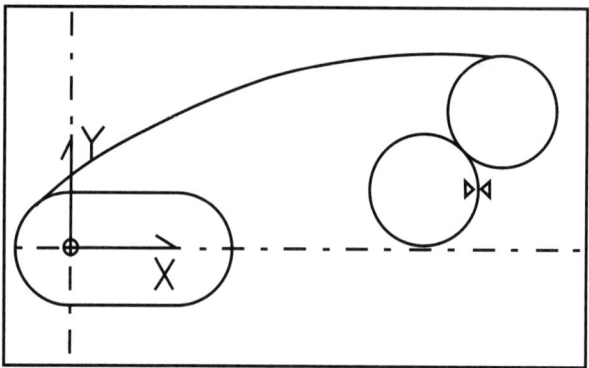

5. Choose **Geometry** from the workbench.
6. Choose **Arc**.
7. Construct a .50″ radius arc at the correct location as per the print.

   **Arc Direction** is counterclockwise.

   **Radius** is .500.

   **Center Point** is X 4.00 - .707, Y 1.25 - .707.

   Choose the **Full Arc** button.

At this time, your process model should match Figure 8.9.

Continue with your process model by constructing the 3.5″ radius arc tangent to the .50″ radius arc which you just constructed and the .530″ radius arc.

Remember, it will also be a counterclockwise arc.

After the completion of the construction of the 3.5″ radius arc, trim the unwanted portions of the .50″ radius arc to match Figure 8.10. Select the **Which Segments** option to view all possible solutions. Remember, with the Trim/Extend function of SmartCAM, pick the portion of the arc you wish to keep.

Continue with your model by trimming and deleting the unwanted geometry until you get a blended profile as shown in Figure 8.11.

It is very probable that, even though the external geometry is complete, it is incorrect. Due to the method that was required to place the geometry, the

**FIGURE 8.10**
The construction of the 3.5″ radius arc

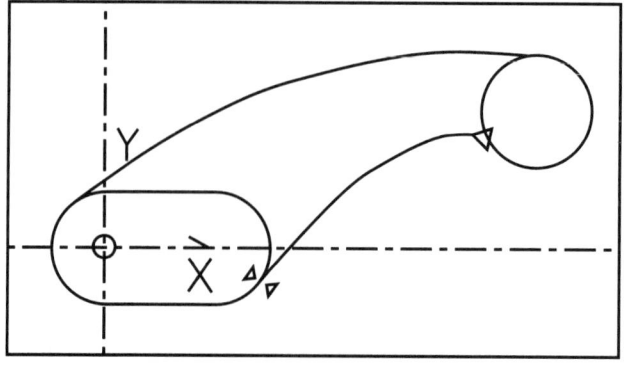

**FIGURE 8.11**
The finished external profile

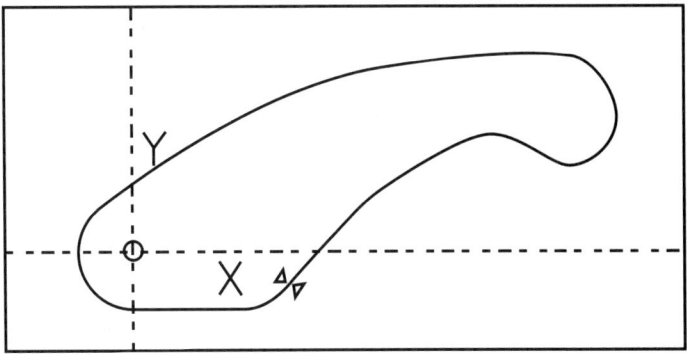

sequence of the database is most likely incorrect. Remember, the result of an out-of-sequence database is out-of-sequence machining. To verify the geometry, simply run the **Show Path** function of SmartCAM.

If the elements are out of sequence, the **Order Path** function will allow you to resequence them.

1. Choose **Edit** from the menu bar.
2. Choose **Order Path**.
3. Choose **Chain**.
4. From the **Chain** control panel, verify the Chain option has been selected.

The chain function will convert connected elements into a true profile. The element which is selected will become the starting element of the profile.

5. Select the straight line of the profile.
6. Run **Show Path** again to verify the corrections.

If, by chance, the elements fail to properly chain, check to verify that all geometry is on the same "Z" level, and that all geometry is connected.

The next objective is to add the lead in/lead out geometry. To do this, split the straight line and insert the lead in/lead out at the split.

1. Choose **Edit** from the menu bar.
2. Choose **Geo Edit**.
3. Choose **Split** from the tool list.
4. Split the straight line at the halfway point.

Now insert the lead in/lead out move.

1. Choose **Lead In/Out** from the Geo Edit tool list.
2. Fill in the input fields of your **Lead In/Out** control panel to match those of Figure 8.12. (Complete the **Change Start**, the **Both**, the **Arc**, the **Angle**, and the **Radius** input fields prior to filling in the **Select Element in Profile** input field.)

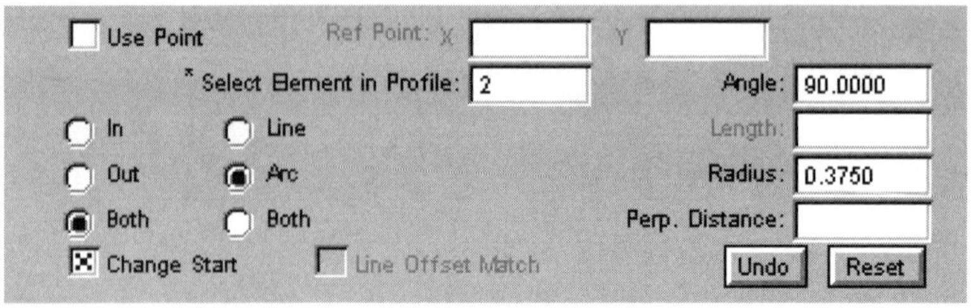

**FIGURE 8.12**
The **Lead In/Out** control panel

Figure 8.12 will insert both a lead in and a lead out arc. Additionally, the starting position of the first element will change so the lead in/lead out will be in the center of the line. An arbitrary angle and radius was chosen. These input fields can be modified to suit individual applications.

At this time, the finish profile is complete and should match Figure 8.13.

## Constructing the Roughing Cuts

The next objective is to create the roughing passes from the finish profile.

To do so, follow these steps:

1. Choose **Group** from the workbench.
2. Choose the **Select All** button.
3. Choose **Insert** from the workbench.
4. Choose the **Before**, **Step Sequence**, and **With Step** options from the tool list.
5. Turn off the **Match Elmt** option.
6. Fill in the input fields of your **Insert** control panel to match those of Figure 8.14.
7. Choose **Geometry** from the workbench.
8. Choose **Wall Offset** from the tool list.

**FIGURE 8.13**
The completed finish profile

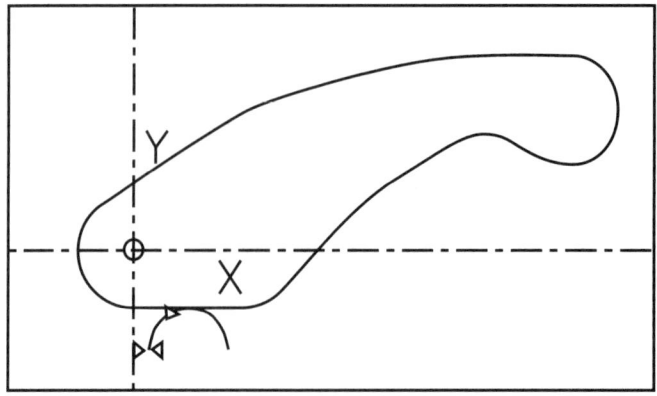

**FIGURE 8.14**
The input fields of the **Insert** control panel

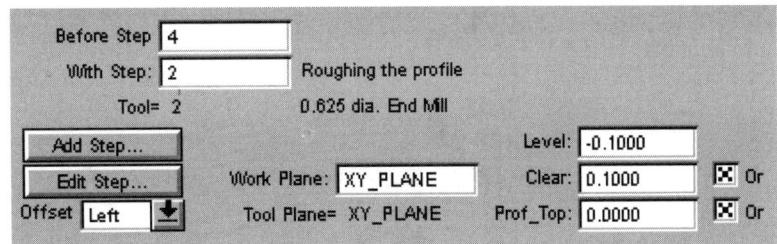

**Wall Side** should be set to left.

**Distance** should be set to .010″.

Choose the **Group Wall** button.

At this time your process model should consist of two profiles: one roughing profile at Z -.100″ and a finish profile at Z - .875″. Your model should match Figure 8.15.

Continue constructing the roughing profiles by transforming copies of the roughing profile down .100″ each.

1. Choose **Group** from the workbench.
2. Choose the **New Group** button to remove any active groups.
3. Choose **Step** from the tool list.
4. Choose step 2, the roughing profile, from the graphics work area.
5. Choose **Edit** from the menu selection.
6. Choose the **Transform** toolbox.
7. Choose **Move** from the tool list.
8. Turn **Copy** on.
9. Set **Copies** to 8.
10. Choose the **From 0** button to fill out the **From Point** input fields.
11. Input X 0.00, Y 0.00, Z -.775/8 in the **To Point** input fields.

At this time, the external roughing passes should be complete and your process model should match Figure 8.16.

**FIGURE 8.15**
The result of the **Wall Offset** function

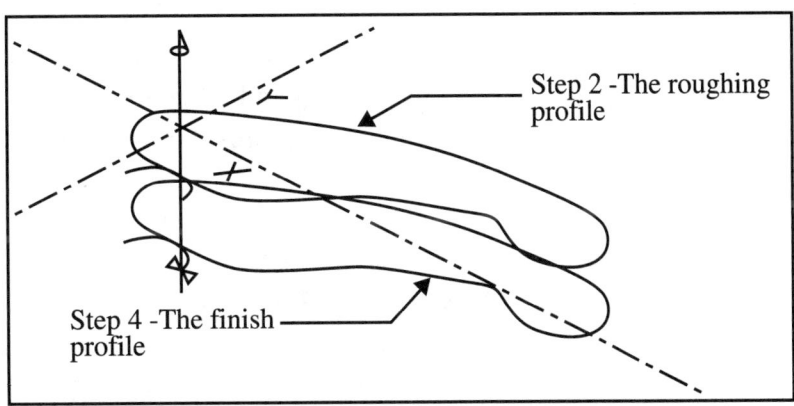

**FIGURE 8.16**
The external roughing and finishing passes

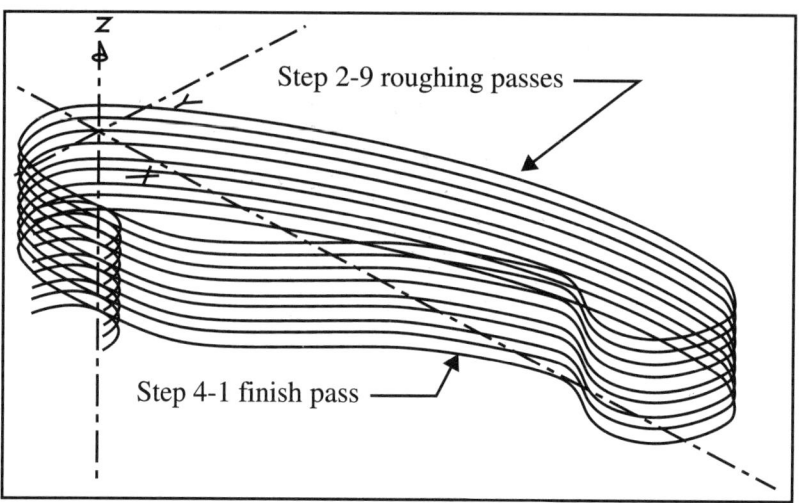

## Machining the Pocket

The pocket roughing routine will be created much like the pocket roughing routine in Tutorial 6, the Support Bracket. The first step is to mask step 2 so your finish profile can be more clearly seen.

1. Choose **Show/Mask** from either the **Utility** menu or choose the icon shown at left.
2. Hide step 2.
3. Choose **Accept**.

Build a pocket boundary on layer 1 at a Z level of -.3125":

1. Choose **Insert** from the workbench.
2. Choose **After** and **Step Sequence** from the tool list.
3. Turn on the **On Layer** selection.

    Set **After Step** to 4.

    Set the **Level** to Z-.3125.

    Set the **Prof Top** to 0.00.

4. Choose **Group** from the workbench.
5. Choose **Step** and group step 4.
6. Choose the **Remove** and the **Element** options from the tool list.
7. Remove the lead in/lead out arcs from the active group. They are not required for the construction of the material boundary.
8. Choose **Geometry.**
9. Choose **Wall Offset**.

    **Wall Side** should be set to left.

    **Distance** should be set to .3125.

    Choose the **Group Wall** button.

**FIGURE 8.17**
The boundary for the pocket roughing routine

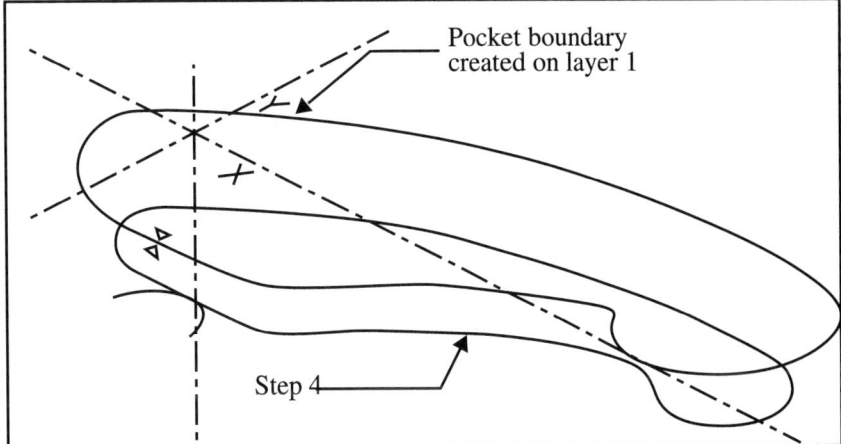

You should now see geometry representing a material boundary at a Z level of -.3125" as shown in Figure 8.17.

Reconstruct the two islands, the elongated island with the .530" arcs and the .50" radius island, which you previously deleted. Both islands should be constructed on layer 1, at a "Z" level of -.3125, and have a Prof Top value of 0.000.

1. To facilitate construction of the islands, obtain a full-scale, top view of the model. In addition, mask step 4, the finish profile.
2. Choose **Geometry**.
3. Choose **Arc**.
4. Create the two arcs with the two tangent lines and then trim them to match Figure 8.4. Remember, the geometry you construct must be a true profile (sequential geometry that is connected) in order to work correctly during the island avoidance portion of the pocket roughing routine.
5. Continue by creating the .50" radius arc as shown in Figure 8.6.

Both island profiles should have a level of -.3125" and a profile top of 0.00. At this time, your process model should match Figure 8.18.

Continue with your process model by constructing a starting point for the pocket roughing routine as was done in the previous tutorial.

1. Choose **Geo Edit**.
2. Choose **Split**.

Inside the **Split** control panel:

Enter a **Gap Width** of .625.

Split the element by the **Element Division** option.

Accept the default .5 percent length.

Select the 6.0" radius arc on the top side of the material boundary geometry.

**FIGURE 8.18**
The pocket material boundary and island profiles

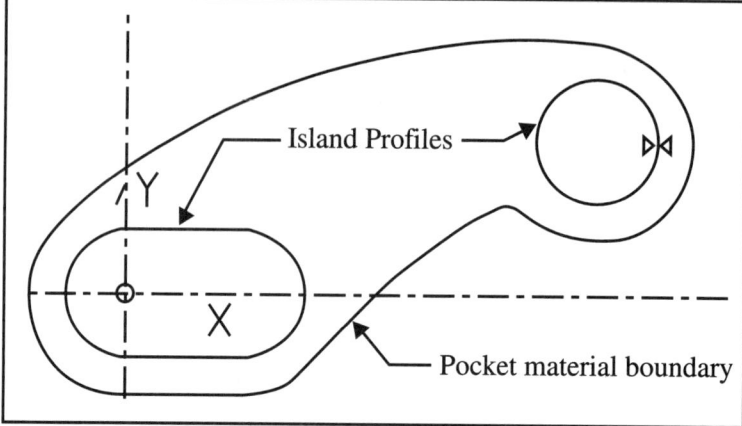

Once the geometry is split, you need to insert a clockwise arc to fill in the gap. To insert a clockwise arc to fill the gap, follow these steps:

1. Choose **Insert**.
2. Choose the **After, Element Sequence,** and **On Layer** options from the tool list.

Inside the **Insert** control panel:

> For **After Element**, choose the left portion of the 6" radius arc.
>
> **Level** is set to -.3125", the pocket depth.
>
> **Prof Top** is set to 0.000".

3. Choose **Geometry**.
4. Choose **Arc**.

> Insert a clockwise arc with a radius of .3125".
>
> As the starting point, choose the left portion of the split arc. Choose the arc close enough to the end so the mouse will snap to the end point. You may need to turn on the proper snap point icons if they are not already on.
>
> As the end point, choose the right portion of the split arc.

At this time, your process model should match Figure 8.19.

Recall from the previous tutorial, there are two factors that control the starting point of the cutting tool during a pocket roughing routine. One is the settings of the Spiral Parameters dialog box and the other is the starting point of the material boundary profile.

To make sure your pocket roughing routine starts at the point where you inserted the arc, you need to make the inserted arc the first element in the boundary profile.

To do so:

1. Choose **Edit** from the menu bar.
2. Choose the **Order Path** toolbox.

**FIGURE 8.19**
The inserted arc

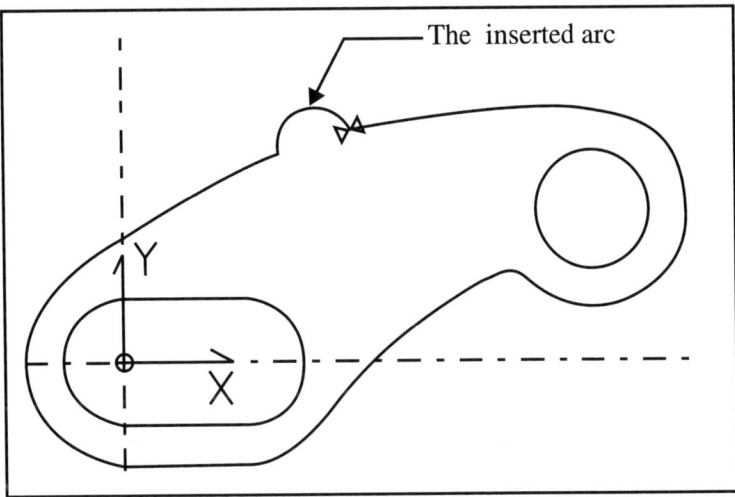

3. Choose **Chain** from the tool list.
4. Inside the **Chain** control panel, verify the **Chain** option is selected.
5. Highlight **Select Element in Profile** by choosing it with your mouse.
6. Select the .3125" arc which you just inserted in the profile. This is now the starting element in the profile.

You are now ready to begin the actual pocket roughing routine.

1. Choose **Insert** from the workbench.
2. Select **After**, **Step Sequence**, and **With Step** from the tool list.
3. Input **After Step** 2 and **With Step** 3 in the input fields of the **Insert** control panel.
4. Set the **Offset** to **None**.
5. The **Level** should be set to 0.00.
6. **Clear** should be set to .100".
7. Set the **Prof Top** value to 0.00.

As in previous tutorials, you will also need to avoid the islands within the pocket. In order to do so, the islands should be the active group. **Place the two islands in the active group before you go to the next step**. (Make sure that both islands are constructed at a level of -.3125" and a profile top of 0.00".)

1. Choose **Process** from the menu bar.
2. Choose the **Rough** toolbox.
3. Choose the **Spiral Parameters** button.
4. Turn the **Cut Inside Out** option off.
5. Choose the **Accept** button.

**FIGURE 8.20**
The input fields of the **Pocket** control panel

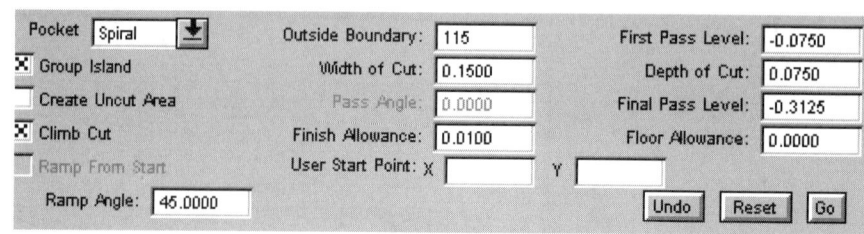

6. Choose the **Pocket** option.
7. Fill in the input fields of your **Pocket** control panel to match those of Figure 8.20.

These values, as with all other values in this text are merely a guide. You may need to change the values to more appropriately meet your needs.

8. Upon choosing the **Go** button, the pocket roughing profiles will be created.

At this time, the pocket roughing routine is complete. Your process model should match Figure 8.21.

Although the pocket roughing routine will remove the majority of the material in the pocket, it will not remove the material where the islands and the material boundary form a corner. This is due to the fact that the pocket roughing routine will not invade the material boundary. This will be easy to see if you will show the toolpath from the top view.

1. Obtain a full-scale, top view of the model.
2. In order to see the finish part profile, show the finish profile with the Show/Mask feature.
3. Run the Show Path feature of SmartCAM.
4. Prior to selecting the Start button, set the Show Tool input field to Filled.

Notice the areas where the material remains uncut in the corners of the pocket. You will need to add roughing and finishing profiles around the islands to correct this problem. You will work with the elongated island first.

5. Hide the finish geometry with the Show/Mask function.

**FIGURE 8.21**
The completed pocket roughing routine

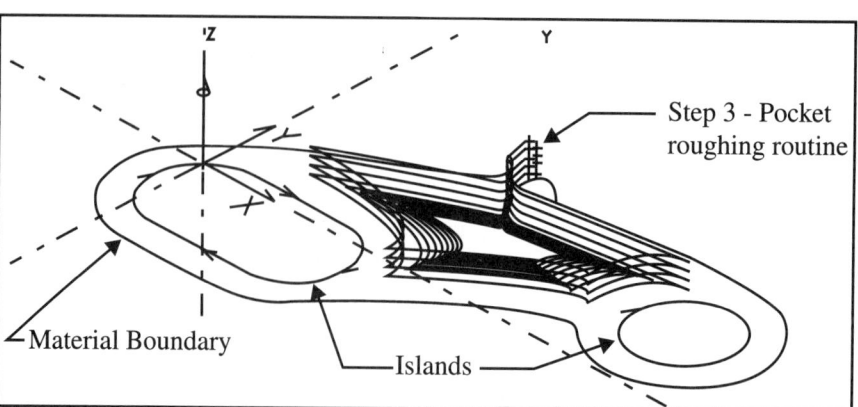

# Chapter 8 SmartCAM Tutorial 7

**FIGURE 8.22**
The geometry which is to be deleted

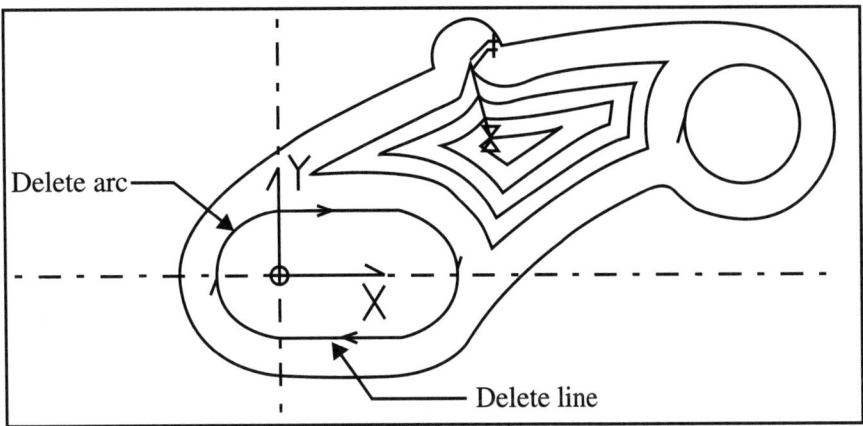

Since the top and the right of the island are the only consideration, you can delete the other portions of the geometry. Figure 8.22 further clarifies the geometry to delete.

After the appropriate geometry has been deleted, trim the remaining arc at approximately the 315 degree mark. To trim the remaining arc:

1. Choose **Geo Edit** from the workbench. If it is not on the workbench it can be found under the **Edit** menu.
2. Choose **Split**.
3. Set **Gap Width** to 0.00.
4. Choose the **Nearest Point** option as the method to split and highlight the **Near Point** input field.
5. Move the cursor to the 315 degree mark and click the left mouse button.
   (Make sure the Free Coordinate Mode is active.)
6. Highlight the **Select Split Element** input field and choose the arc.

This will split the arc into two segments.

7. Delete the bottom portion of the arc.

The modified arc should now match Figure 8.23.

Next, the 1.0" arc should be trimmed so that only that portion of the arc from 90 degrees to 270 degrees is left. To trim the 1.0" arc:

1. Choose **Geo Edit** from the workbench.
2. Choose **Split**.
3. Choose the **Nearest Point** option.
4. Split the line twice, once at the 90 degree point and once at the 270 degree point.
5. Delete the right portion of the arc.

The remainder of the arc should match Figure 8.24.

**FIGURE 8.23**
The modified arc of the elongated boss

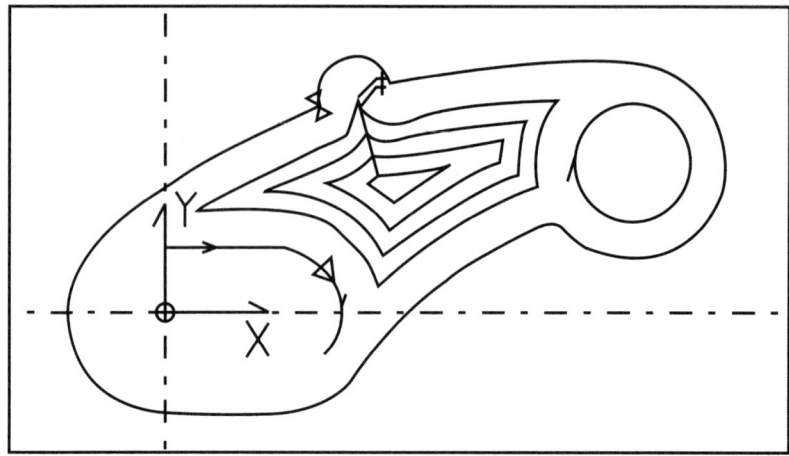

**FIGURE 8.24**
The modified arc of the 1″ diameter boss

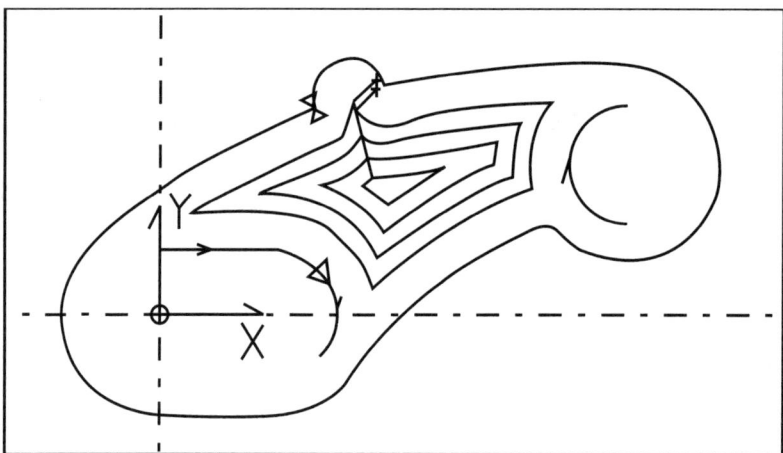

The next step will change the three grouped elements on layer 1 into the finish tool. You must, however, show the finish tool before you complete the next step.

***Remember:*** *Show the finish geometry before proceeding to the next step.*

To change the grouped geometry into the finish toolpath:

1. Choose **Edit** from the menu bar.
2. Select **Property Change.**
3. Choose **Tool Path**.
4. Fill in the input fields of your **Toolpath Property Change** dialog box to match those of Figure 8.25.

Both the **Clear** and the **Prof Top** input fields must be turned on before values can be entered.

You will next add lead in/lead out moves to both of these profiles.

Chapter 8 SmartCAM Tutorial 7

**FIGURE 8.25**
The input fields of the **Toolpath Property Change** dialog box

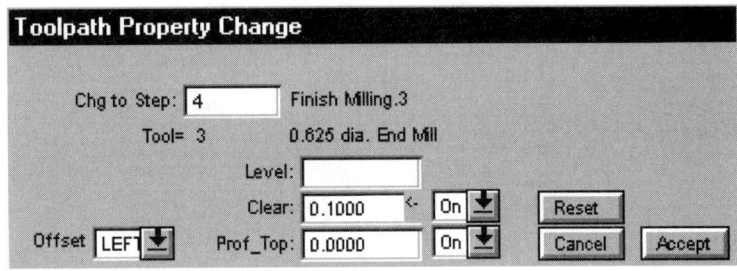

1. Choose **Geo Edit** from the workbench.
2. Choose **Lead In/Out** from the tool list.
3. Choose **Both** a lead in and a lead out.
4. Choose **Arc**.
5. Input a 90 degree arc of .50″ radius
6. Highlight **Select Element in Profile** and choose both profiles.

A lead in/lead out arc should now appear on both profiles as shown in Figure 8.26.

Since you just completed the finish passes for the islands, you now need to insert the roughing passes.

1. Choose **Insert** from the workbench.
2. Select the **After** and the **Step Sequence** options.
3. Choose the **With Step** option.

Fill in the input fields of your **Insert** control panel to match those of Figure 8.27.

1. Choose **Geometry** from the workbench. If **Geometry** is not on the workbench it may be found under the **Create** menu.
2. Choose **Wall Offset**.

**FIGURE 8.26**
The inserted lead in/lead out arcs

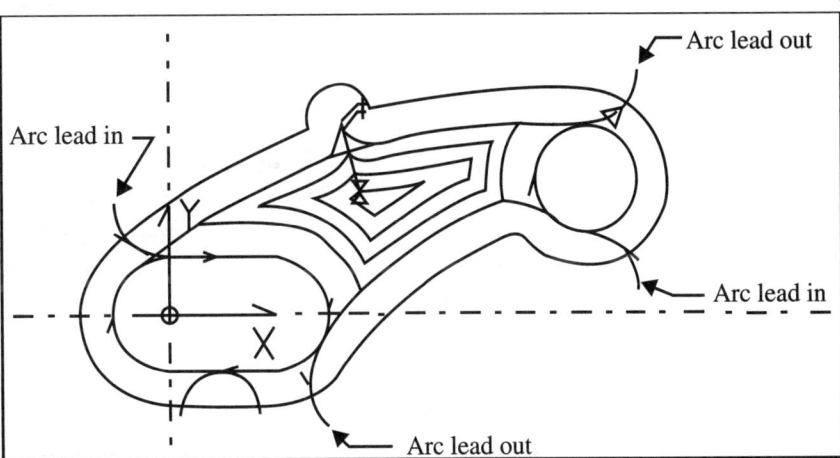

**FIGURE 8.27**
The input fields of the **Insert** control panel

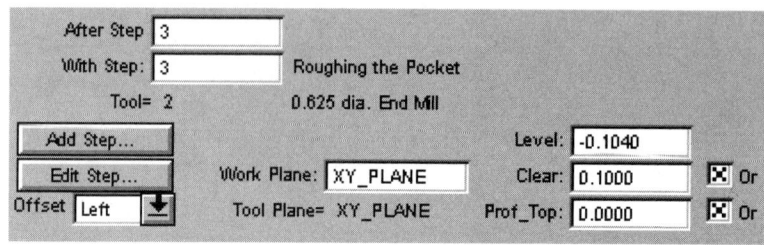

3. Insert a left offset of .010".
4. Choose the **Group Wall** button.
5. Choose **Group** from the workbench.
6. Choose the **New Group** button.
7. Select the **Profile** option.
8. Group the roughing profiles which were just created by means of the **Wall Offset**.
9. **Transform** the profiles down to create two additional copies at an incremental distance of .104" each.

This will complete the roughing and finishing passes on the islands. At this time, your process model should match Figure 8.28.

# Spot Drilling and Drilling the Holes

The print calls for one .50" hole in the part. This hole will be drilled with a 1/2" drill. However, you will also drill two holes in the slot in order to provide clearance for the 1/2" endmill.

1. Choose **Insert** from the workbench.
2. Choose **After**, **Step Sequence,** and **With Step** from the tool list.

**FIGURE 8.28**
The completed roughing profiles of the islands

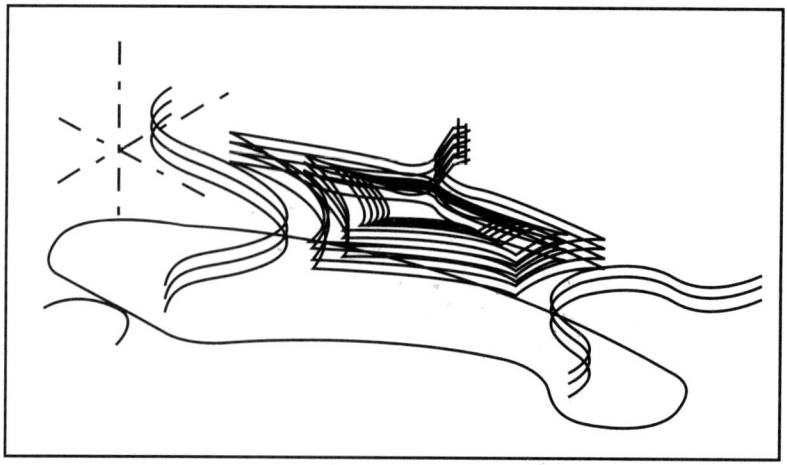

# Chapter 8  SmartCAM Tutorial 7

3. Input **After Step 4** and **With Step 7.**
4. Input an **Offset** of None, a **Level** of 0.00, a **Clear** of .100, and a **Prof Top** of 0.00.
5. Choose **Geometry** from the workbench.
6. Choose **Hole** from the tool list.
7. Inside the **Hole** control panel, input a **Spot Diameter** of .375.
8. For the **Hole Point** input field, input the following locations in succession:

    Location 1 - X 0.00 Y 0.00
    Location 2 - X 1.00 Y 0.00
    Location 3 - X 4.00 Y 1.25

You should see three points representing the three spot drilled holes.

Continue with your process model by placing the three drilled holes according to the following instructions:

1. Choose **Insert** from the workbench.
2. Insert the drill, step 8, after the spot drill, step 7.
3. Verify the Level is set to 0.000.
4. Accept the default for all other fields.
5. Choose **Geometry** from the workbench and **Hole** from the tool list.
6. Input a **Full Depth** of .875.
7. Highlight the **Hole Point** input field and input the following locations in succession:

    Location 1 - X 0.00 Y 0.00
    Location 2 - X 1.00 Y 0.00
    Location 3 - X 4.00 Y 1.25

The isometric view of the model will clearly show the representations of the three drilled holes that were just created.

## Machining the Slot

The next step in the construction of your process model is roughing and then finishing the .56" by 1.00" slot.

1. Obtain a full-scale, top view of the model.
2. Choose **Insert** from the workbench.
3. Choose **After, Step Sequence,** and **With Step** from the tool list.
4. Input **After Step 8, With Step 6.**
5. Input a left **Offset**, a **Level** of -.875, a **Clear** of .100, and a **Prof Top** of 0.00.
6. Choose **Geometry** from the workbench.

7. Choose **Arc** from the tool list.
8. Complete the input fields of your **Arc** control panel to match those of Figure 8.29.

After completion of the **Arc Direction** input field, the **Center Point** input field and the **Radius** input field, choose the **Full Arc** button.

9. Highlight the Center Point input field and input an "X" value of 1.00 and a "Y" value of 0.00.
10. Construct the two lines that will complete the slot. Remember to construct the lines in a counterclockwise manner. Use the **Tan Arc** method of constructing the lines.
11. Trim the two arcs, as was demonstrated previously to form the complete profile.
12. Sequence the elements to form a true profile:

    Choose **Edit**.
    Choose **Order Path**.
    Choose **Chain**.
    Select the bottom line as the starting point of the profile.

13. Add a 90 degree, .25" radius lead in/lead out move which will allow the tool to plunge into the hole which was previously drilled in the slot.

The construction of this slot is identical to the construction techniques used to create the elongated island that was previously demonstrated in this tutorial. You may need to review the techniques used in order to correctly construct the slot.

Figure 8.30 shows the slot, complete with lead in/lead out arcs. All irrelevant geometry has been masked to clarify the view.

In order to rough the slot, follow these steps:

1. Choose **Group** from the workbench.
2. Select the **New Group** button to clear the active group.
3. Select the **Profile** option from the tool list.
4. Select the profile which defines the slot.
5. Select **Insert** from the workbench.
6. Choose **Before**, **Step Sequence**, and **With Step** from the tool list.
7. Insert step 5 before step 6.

**FIGURE 8.29**
The input fields of the **Arc** control panel

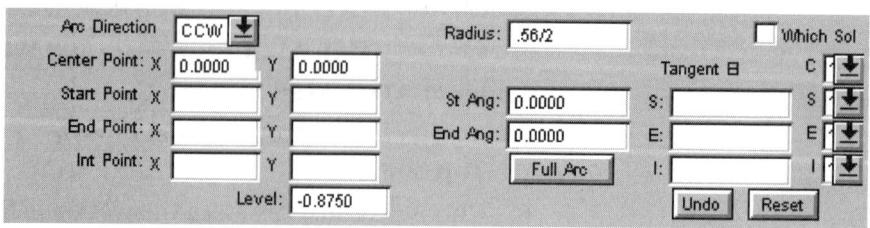

**FIGURE 8.30**
The completed slot profile

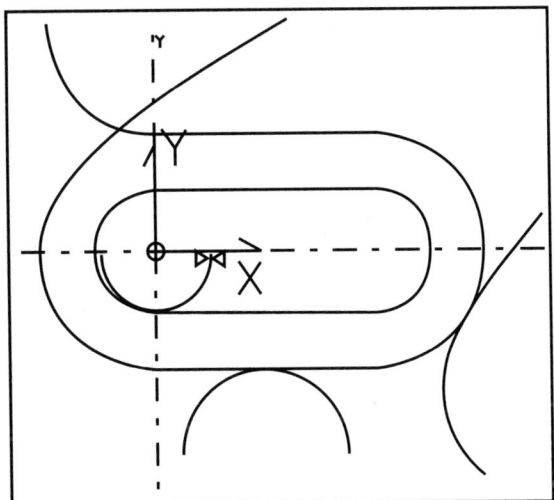

8. Choose a left **Offset**.
9. Input a **Level** of -.100, a **Clear** of .100, and a **Profile Top** of 0.00.
10. Choose **Geometry** from the workbench.
11. Choose the **Wall Offset** tool from the tool list.
    Set **Wall Side** to Left.
    **Distance** is set to .010.
    Choose the **Group Wall** button.

Geometry representing the roughing pass should now be shown in your graphics work area as shown in Figure 8.31.

Continue creating the roughing geometry by grouping the first roughing profile and then transforming the profile down to the specified depth:

**FIGURE 8.31**
The slot roughing and finishing profiles

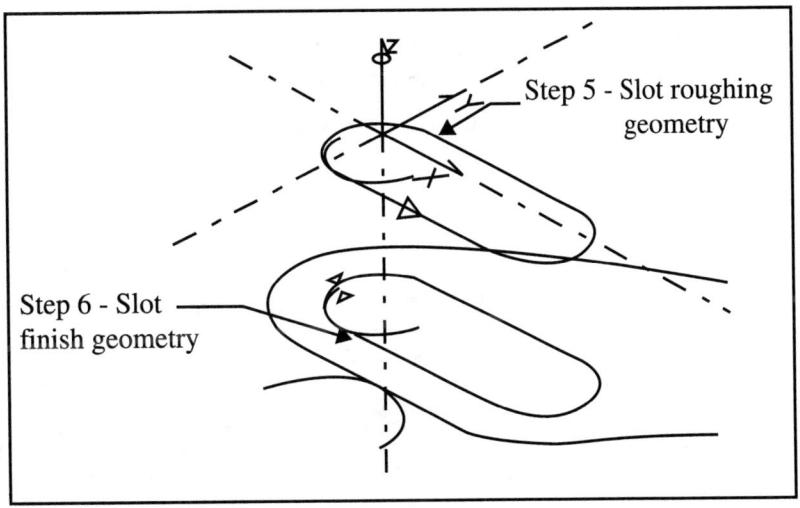

1. Choose **Group** from the workbench.
2. Choose the **New Group** button and the **Profile** option from the tool list.
3. Select the first roughing profile of the slot in the graphics work area.
4. Choose **Edit** from the menu bar.
5. Choose the **Transform** tool box and the **Move** tool.
6. Fill in the input fields of the **Move** control panel according to Figure 8.32.

At this time, the roughing and finishing passes of the slot should be complete.

## Facing the Workpiece

Your next objective is to face the workpiece to the required thickness. In an attempt to remain consistent with industrial procedures, assume the blank material is 1.00″ in thickness and you need to face an equal amount of material from each side. The first facing pass will take place at a "Z" level of 0.00 (the actual material removal will require adjustment of the G54 "Z" by .0625″). You will then flip the work and face .0625″ from the other side. Additionally, the insertion of user events will be required to complete this task.

Before beginning the next sequence of steps, hide steps 2 through 8 and show layer 1, the material boundary for the pocket roughing routine. This will make the graphics work area less congested with toolpath and therefore easier to see the geometry that represents the facemill. Additionally, you will need to view the model from the top to more easily see the processes.

1. Choose **Insert** from the workbench.
2. Choose the **Before, Step Sequence,** and the **With Step** options from the tool list.
3. Insert step 1, the facemill, before step 2.
4. Set the **Offset** to none, the **Level** to 0.00, the **Clear** to .100″, and the **Prof Top** to 0.00.
5. Choose **Process** from the menu bar.
6. Choose the **Rough** toolbox.

**FIGURE 8.32**
The input fields of the **Move** control panel

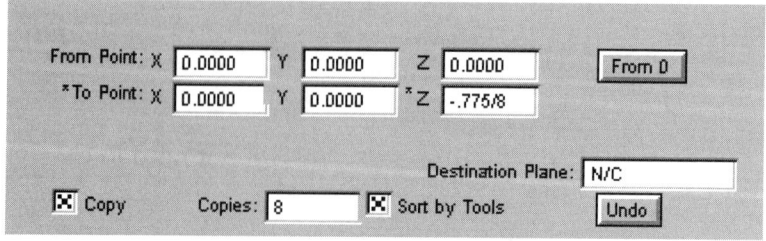

**FIGURE 8.33**
The input fields of the **Face** control panel

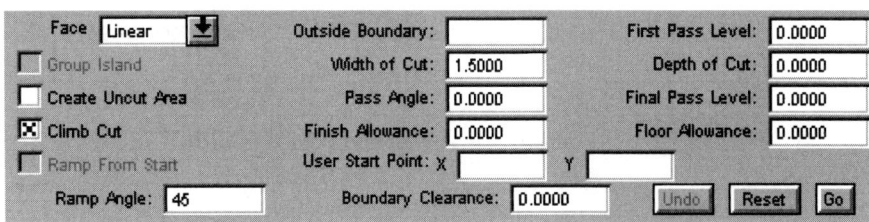

7. Choose the **Face** tool.
8. Fill in the input fields of your **Face** control panel to match those of Figure 8.33.

(The Outside Boundary input field was intentionally left blank in this Figure. To input a value for this input field, simply highlight the input field and select the profile that was constructed on layer 1.)

9. Choose the **Go** button.

At this time you should see linear roughing passes covering your process model.

As was explained in the previous tutorial, geometry may be created which is unnecessary. To maximize production, simply delete that portion of the facing pass that is not needed.

At the time of the actual machining cycle, you would now need to flip the part in the vise in order to machine the back side of the part. This will require the insertion of user events.

1. Choose **Create** from the menu bar.
2. Choose the **User Elemts** toolbox.
3. Choose the **User Event** tool.
4. Input "G28 M00" into the **Event Text** input field.
5. **Location Point** can be set to "X" 0.00 and "Y" 0.00.
6. **Level** is set to 0.00 also.
7. Select the **Go** button to insert this event.

Additional user events will be required to restart the spindle, the coolant, and other miscellaneous functions. Enter the appropriate commands according to your machine tool.

Next, you will need to transform the facing geometry down an additional .0625" to face the remaining material from the blank.

1. Choose **Group** from the workbench.
2. Choose **New Group** and **Step** from the tool list.
3. Select step 1, the facemill.
4. Select the **Remove** option and remove the User Event from the active group. We do not need to copy this element.
5. Choose **Transform** from the workbench. (If Transform is not on the workbench it is found under the Edit menu.)

6. Select the **Move** option from the tool list.
7. Copy the grouped geometry down an additional .0625″.

At this time the construction of the facing step is complete.

## Constructing Additional Workplanes

Additional workplanes are required to complete the machining processes in this tutorial also. You will simply duplicate the steps that were done in the previous tutorial to construct the workplanes of this tutorial.

1. Choose **Workplane** from the menu bar.
2. Choose the **Define Plane** toolbox.
3. Fill in the input fields of your **Define Plane** control panel to match those of Figure 8.34.
4. Choose the **Accept** button.

Continue your process model by constructing another workplane. Name the workplane "G56." All other input fields will be the same. Recall that once you define a workplane, it becomes the active workplane. Therefore, even though the "G56" workplane has the same values as the "G55" workplane, the "G55" values are from the "XY" workplane and the "G56" values are from the "G55" workplane.

Construct the "G56" workplane before you proceed with the tutorial.

1. Choose **View** from the menu bar.
2. Choose **Zoom**.
3. Input a **Zoom Magnification Factor** of .5.
4. Choose the right side of the process model as the **View Center**.
5. Choose **Insert** from the workbench.
6. Inside the **Insert** control panel, toggle back and forth between the "G55" and the "G56" workplanes to verify that both workplanes were properly constructed.

At this time, your process model should consist of three workplanes, the original "XY" workplane, the "G55" workplane and the "G56" workplane. All geometry, however, should still reside on the "XY" workplane.

**FIGURE 8.34**
The definitions for the G55 workplane

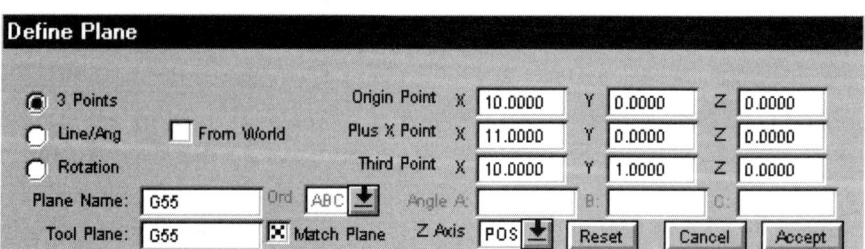

# Chapter 8  SmartCAM Tutorial 7

The next objective in the construction of your process model is to place the proper geometry on the proper workplane. Step 1, the face mill, step 7, the spot drill, and step 8, the drill will reside on the "XY" workplane. Additionally, the slot will be roughed and finished on the "XY" workplane. The workpiece will then be moved to the "G55" workplane and the complete profile will be roughed and finished. Additionally, the pocket on the top of the work will be roughed and finished at the "G55" workplane. The part will then be flipped and the pocket on the bottom of the workpiece will be roughed and finished.

To move the geometry to the proper workplane:

1. Select the **Show/Mask** icon and show all steps and layers
2. Choose **Group** from the workbench.
3. Select the **New Group** button.
4. Choose the **Step** option.
5. Choose steps 2, 3, and 4.
6. Choose **Transform**.
7. Choose **Move** (make sure the Copy option is off).
8. Inside the **Move** control panel, choose **Destination Plane** and choose "G55" from the data list.

The geometry that represents the profile and pocket roughing and finishing cuts should move to the "G55" workplane.

To copy the appropriate geometry to the "G56" workplane:

1. Select the **Show/Mask** icon and hide step 2, the profile roughing step.
2. Choose **Group** from the workbench.
3. Choose the **Remove** and the **Profile** option from the tool list.
4. Choose the geometry which represents the external finish profile as shown in Figure 8.35.
5. Choose **Transform**, **Move**.

**FIGURE 8.35**
The profile which is to be removed from the active group

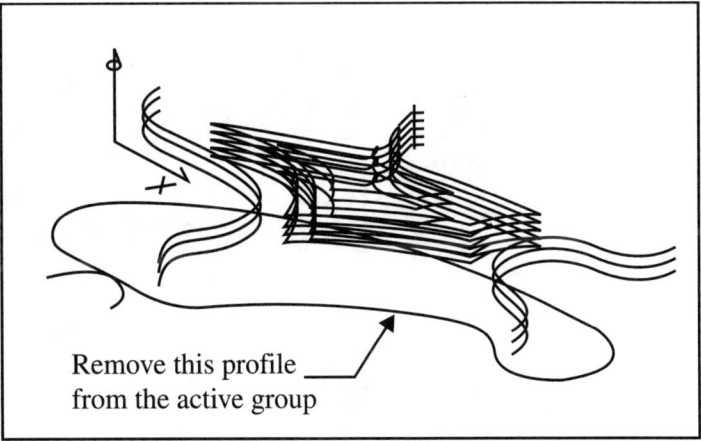

6. Inside the **Move** control panel, turn the **Copy** option on and set **Copies** to 1.
7. Again, choose **Destination Plane** and choose "G56" from the data list.

Geometry that represents the pocket roughing and finishing passes should now be copied to the "G56" workplane.

Figures 8.36, 8.37, and 8.38 should further explain the previous sequence of events.

Once all geometry is located on the proper workplanes, you will need to insert the appropriate user events to facilitate the moving of the workpiece to the different workplanes.

1. Choose Show/Mask and show all steps.
2. Choose **Insert** from the workbench.

**FIGURE 8.36**
The G54 (XY) workplane

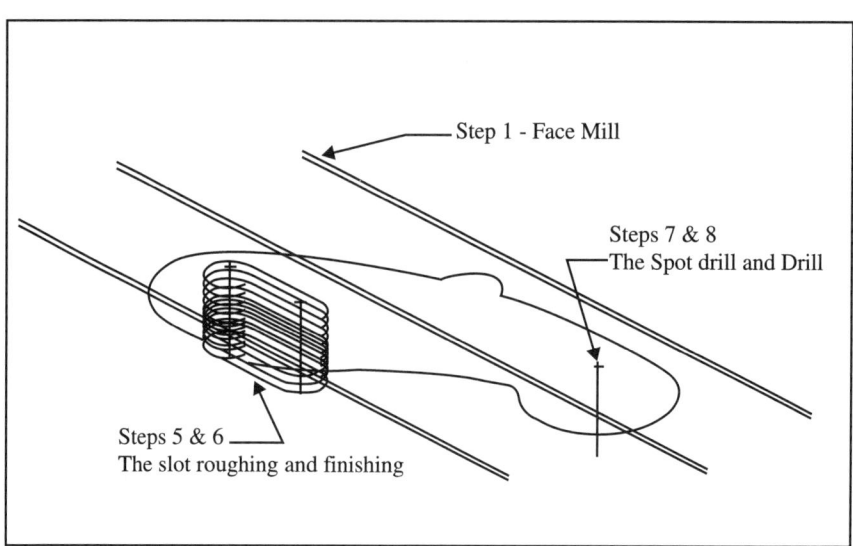

**FIGURE 8.37**
The G55 workplane

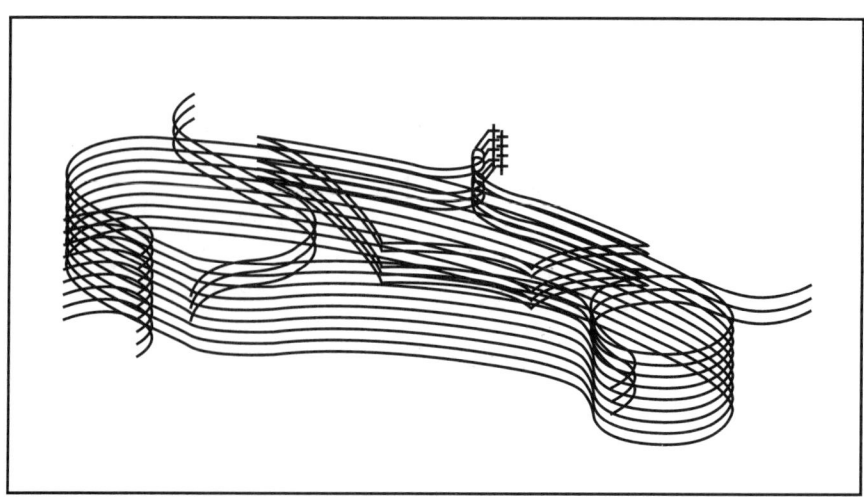

Chapter 8   SmartCAM Tutorial 7

**FIGURE 8.38**
The G56 workplane

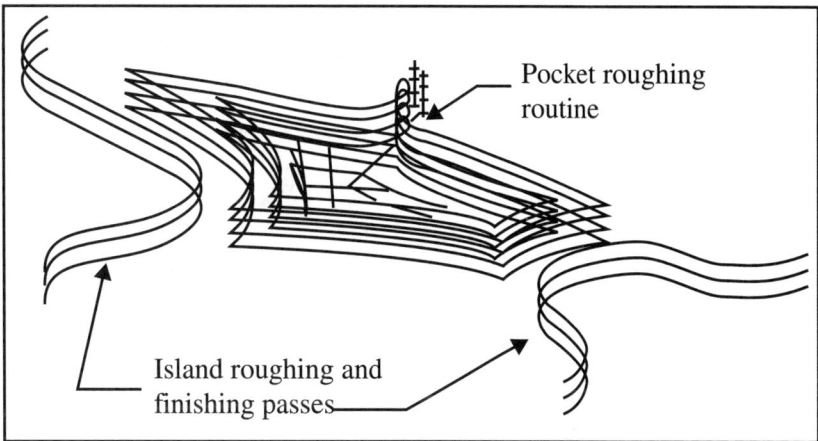

3. Choose the **Before**, **Step Sequence,** and the **With Step** options from the tool list.
4. Input **Before Step** 2, **With Step** 2.
5. Choose the **Workplane** input field from within the **Insert** control panel.
6. Choose the G55 workplane from the data list.
7. Accept the default for all other input fields.
8. Choose **Create** from the menu bar.
9. Choose the **User Elmts** toolbox.
10. Choose the **User Event** tool.
11. Insert the following user events into your database:

    G28 M00 (this will move the cutting tool to a safe location to allow the operator to move the workpiece to the G55 workplane)

    S512 M03 (turn the spindle back on after the M00)

    M08 (turn the coolant back on)

    G00 G55 (set the proper workplane)

Additional codes may be entered as you deem appropriate.

Additional user elements will need to be inserted before the third workplane as well. Insert the appropriate user elements on the "G56" workplane as was done in the previous explanation.

Run the **Show Path** function of SmartCAM and very carefully watch to see the sequence of events. It is quite possible that the machining will be done out of sequence due to the shifting of the geometry to the different workplanes.

To properly sequence the machining:

1. Choose **Group** from the workbench.
2. Select the **New Group** button.
3. Choose the **Step** option.

4. In this order, choose steps 1, 7, 8, 5, 6, 2, 3, & 4.
5. Inside the **Group** control panel, select the **Sequence Move** button (make sure the **By Selection Order** selector switch is on).
6. Choose the **New Group** button.
7. Change the group selection method to **Box** and box in the entire G55 workplane.
8. Box in the entire G56 workplane.
9. Inside the **Group** control panel, select the **Sequence Move** button (make sure the **By Selection Order** selector switch is on).

Again, run the Show Path function of SmartCAM to verify the correct sequence of machining.

This completes the final tutorial of this text.

# Index

Arc control panel, 49, 70–71
   tangent element, 87, 115, 173

Blend. *See* Geo Edit

Calculator, 146, 154
Chamfer. *See* Geo Edit
Control panel, 9. *See also* Arc; Face; Geo Edit; Hole; Insert; Line; Pocket Roughing; Transform; Wall Offset

Database list, 9
Delete, 115
Dialog box, 4
Display Modes dialog box, 15

Edit Filter, 11–12, 13–14
Element Data, 91, 93, 94, 114, 119
Environment. *See* SmartCAM environment

Face control panel, 105, 161, 191
Free coordinate mode, 12–13

Geo Edit, 5
   blend, 65–66, 111, 117
   chamfer, 66–67
      control panel, 67
   lead in/lead out, 68–69, 77–78, 95–96, 112,
      control panel, 68
   split, 95–96, 111
      gap width, 155
Graphics work area, 15
Group toolbox, 7–8
   group tool palette, 10–11, 18
   new group, 8, 57, 73, 89
Grouping, 7

Hole control panel, 78–80, 101, 123–124

Icon bar, 4–5
Input field, 9
Insert, 5
   control panel, 9, 44–45, 58, 64, 70
   insert property bar, 16

197

match element, 14, 94–95, 115, 139

Job Operations File, 21
Job Operation Planner, 22
    add button, 24, 38
    add process step window, 24, 38, 42
    choose tool button, 31, 41
    duplicate button, 30
    edit process step window, 24–25, 39–40, 43
    general page, 22, 35, 36
    job information button, 22–23, 36
    machine page, 22, 35, 37
    material page, 23
    move, 29
    operation page, 26, 39, 42, 43
    overview, 20–22
    process planner icon, 22, 35
    remove, 27
    renumber, 29
    tool page, 24–25

Layer, 20, 104, 147, 148
Lead in/lead out. *See* Geo Edit
Line control panel, 45, 48
    tan arc, 50

Match element. *See* Insert
Menu bar, 2
Mirror image. *See* Transform
Move control panel. *See* Transform

New Group button. *See* Group toolbox

Order Path, 91
    chain, 91, 118–119, 143, 156, 175, 180–181
    reverse order, 93, 114

Point/Rapid, 128
Pocket Roughing, 75–77, 99–100, 121–122, 150
    control panel, 75
    group island input field, 150

    spiral parameters input field, 156–157
    user start point input field, 157
Process planner icon. *See* Job Operation Planner
Profile, 8, 179
Profile Top, 147
Property Change, 94
    holes/points, 103–104
    toolpath, 94, 114, 143

Readout line, 15

Scale. *See* Transform
Sequence move, 196
Show Cut, 20, 25
Show/Mask, 103, 104, 178
Show Path, 25, 118, 195
SmartCAM environment, 2, 3
Snap Icons, 10, 13, 65
Snap Mode, 13
    automatic mode, 45
Step. *See* With Step
Submenu, 3

Title bar, 2
Toolbox, 4
    workbench resident toolbox, 6
Tool list, 6
    modeling tools, 7
Transform, 57
    mirror image, 88
        suppress planes, 89
    move, 57, 59–60, 74, 87
        control panel, 74
        from 0 button, 177
    rotate, 89–90, 102
    scale, 92
Trim/Extend, 52–54
    rules for using, 117, 141, 174
    which segments, 90, 141, 174

Unique tool number, 29–30, 40–41
User Event, 127–129, 162, 164, 195

Wall Offset, 56, 58–59, 72–73
With Step, 20, 55
Workbench, 6

Workplanes, 129
  active workplane, 131
  construction of, 130, 163, 192
  define plane dialog box, 130, 163, 192
  explanation of, 129
  reserved workplane, 131
  system workplane, 129